DESTINATION MARS

DESTINATION MARS

The Story of Our Quest to Conquer the Red Planet

ANDREW MAY

ICON

Published in the UK in 2017
by Icon Books Ltd, Omnibus Business Centre,
39–41 North Road, London N7 9DP
email: info@iconbooks.com
www.iconbooks.com

Sold in the UK, Europe and Asia
by Faber & Faber Ltd, Bloomsbury House,
74–77 Great Russell Street,
London WC1B 3DA or their agents

Distributed in the UK, Europe and Asia
by Grantham Book Services,
Trent Road, Grantham NG31 7XQ

Distributed in the USA
by Publishers Group West,
1700 Fourth Street, Berkeley, CA 94710

Distributed in Australia and New Zealand
by Allen & Unwin Pty Ltd,
PO Box 8500, 83 Alexander Street,
Crows Nest, NSW 2065

Distributed in South Africa
by Jonathan Ball, Office B4, The District,
41 Sir Lowry Road, Woodstock 7925

Distributed in India by Penguin Books India,
7th Floor, Infinity Tower – C, DLF Cyber City,
Gurgaon 122002, Haryana

Distributed in Canada by Publishers Group Canada,
76 Stafford Street, Unit 300
Toronto, Ontario M6J 2S1

ISBN: 978-178578-225-1

Typeset in Iowan by Marie Doherty

Printed and bound in the UK
by Clays Ltd, St Ives plc

CONTENTS

PREFACE: FROM SCIENCE FICTION TO SCIENCE FACT

It's the year 2031. The six astronauts making up the crew of Ares 1 blast off into Earth orbit, where they proceed to link up with the recently completed Hermes Mars transit vehicle. Looking like a small-scale version of the ISS – the International Space Station – Hermes boasts something the ISS never had: a nuclear-powered ion engine. Nothing like as powerful as the chemical rocket that lifted the crew off the Earth's surface, this is only capable of producing a tiny acceleration … but it's an acceleration that can be kept up for months on end. That's enough to take Hermes all the way to Mars, putting it safely into orbit around the Red Planet later the same year.

All six crew members transfer to the small Mars descent vehicle, which takes them down to the surface. After entering the thin Martian atmosphere, their descent is slowed first by a large parachute – like a Soyuz space capsule returning to Earth – and then by a powerful downward-pointing rocket engine, like a scaled-up version of the one on the Apollo lunar lander. The Ares 1 crew don't just land anywhere, but in a carefully prepared spot. A series of earlier, uncrewed missions has already delivered everything they need for their

30-day stay, including a pressurised habitat with food, air and water, two surface rovers and a variety of scientific instruments. Most important of all, ready and waiting for them when they arrive is the Mars ascent vehicle, which has been busy making fuel for the trip back up to Hermes by mixing hydrogen with carbon dioxide from the Martian atmosphere.

This first human journey to Mars passes without a hitch, and the Ares 1 crew return safely to Earth – to be followed two years later by Ares 2 on a similar mission. Then after another two years … well, perhaps Ares 3 won't go quite according to plan. It doesn't in Andy Weir's 2014 novel *The Martian*, which was turned into a blockbuster movie by Ridley Scott the following year. That's the source of the scenario that's just been described – and for Weir and Scott it serves to set up one of the most exciting survival stories of modern times. But is it really just fiction?

Most of the technology portrayed in *The Martian* already exists. The Saturn V rocket of the 1960s could lift more than a hundred tonnes of payload into Earth orbit, and there are several launchers in development which will be able to match that. The 400-tonne ISS shows that it's possible to build large-scale structures in Earth orbit – and that humans can safely live and work in space for a year or more.

The huge, nuclear-powered Hermes spaceship seen in the movie is far more sophisticated than it needs to be (for a trip to Mars, that is – it's about right for a Hollywood blockbuster). The issue here, though, is one of unnecessary cost and extravagance rather than lack of technological feasibility. It would certainly be possible, for example, to produce artificial gravity by rotating parts of the spacecraft – but whether that's necessary or cost-effective, on a mission designed to last little more than a year, is a different matter.

Similarly, the 'ion drive' propulsion system – identified in the novel as being of the VASIMR (Variable Specific Impulse Magnetoplasma Rocket*) type – is technically feasible, and the subject of current research, but it's probably unnecessary. Conventional rockets should be able to do the job well enough.

The crew-carrying Hermes would be preceded by various supplies sent in advance – robotic missions of the kind that are already routinely sent to Mars whenever a launch window opens up. This 'split-mission' approach was promoted in the 1990s by aerospace engineer Robert Zubrin, under the title Mars Direct – which also included the idea of in situ fuel production, as portrayed in *The Martian*. With various permutations, Zubrin's basic mission architecture is echoed in most current proposals for the human exploration of Mars, whether from government agencies like NASA or from private companies like Elon Musk's SpaceX – and of course in fictional portrayals like *The Martian*.

So a journey to Mars is technically feasible – but is it going to happen in the next 20 years? During the 1960s, many people were confident there would be humans on Mars before the end of the 20th century. Sadly, that didn't happen, for a variety of reasons, ranging from money and politics to conflicting priorities within the space and science communities. Despite that, progress has been made. We've got the football-field-sized ISS, maintaining a continuous human presence in space since the first year of this century. We've seen NASA put the Curiosity rover – a mobile science laboratory the size of a small city car – on to the surface of

---

* A list of abbreviations is included at the end of the book (see page 155).

Mars, and then drive it up the side of a mountain. And we've heard big-business entrepreneurs like Musk making bullish predictions about reaching the Red Planet on private funds alone. The race to Mars may have had a slow start, but it's definitely hotting up now ...

THE LURE OF THE RED PLANET

1

Our Solar System neighbour

> May we attribute to the colour of the herbage and plants, which no doubt clothe the plains of Mars, the characteristic hue of that planet, which is noticeable by the naked eye, and which led the ancients to personify it as a warrior? Are the meadows, the forests, and the fields, on Mars, all red? … The land cannot be all over bare of vegetation, like the sands of the Sahara. It is very probably covered with a vegetation of some kind, and, as the only colour we perceive on Mars's terra firma is red, we conclude that Martian vegetation is of that colour.

So wrote the French astronomer Camille Flammarion in 1873. That was a time when telescope technology had just got to the point of discerning a few surface features on the Red Planet – and people like Flammarion let their imaginations run wild. Their writings sparked a worldwide surge of interest in Mars, and before long its alleged 'red vegetation' was joined by a complex system of artificial canals and malev-olent bug-eyed monsters. It was just the latest phase in humanity's fascination with the Red Planet, which stretches back to antiquity.

Mars is one of Earth's closest neighbours in space, the next planet out from the Sun. Before the invention of the telescope, it was one of just five planets that could be seen in the night sky. The others are Mercury and Venus, both orbiting closer to the Sun than Earth, and Jupiter and Saturn out beyond Mars. Only Venus ever gets closer to the Earth than Mars.

The Earth's orbit takes it all the way round the Sun once a year. The orbit isn't a perfect circle, but it's close enough, with an average radius of about 150 million kilometres. For convenience, this distance is sometimes referred to as an 'astronomical unit', or AU, to make it easier to compare dis-tances within the Solar System. Venus, for example, travels around the Sun on a near-circular orbit of 0.72 AU radius, so its closest approach to Earth is just 0.28 AU, or about 42 million kilometres.

The situation with Mars is more complicated. Its orbit is distinctly non-circular – an oval shape called an ellipse. Its distance to the Sun at perihelion, or closest approach, is only about four-fifths of that at aphelion, the furthest point of its orbit. Actually, all the planets move on elliptical orbits – that's Kepler's first law of planetary motion – but in the case of Earth and Venus, perihelion is only slightly closer to the Sun than aphelion.

On average, Mars is about 50 per cent further from the Sun than Earth. It also takes significantly longer to complete an orbit, following the third of Kepler's laws which states that a planet's orbital period increases with its distance from the Sun. A Martian 'year' is 687 days long, or just a few weeks short of two Earth years. During this time, the distance between Mars and the Sun varies between 1.38 AU at perihelion and 1.67 AU at aphelion.

Due to a combination of its orbital eccentricity and lack of synchronisation between the Martian year and our own, the distance to Mars – and hence its appearance in the skies of Earth – fluctuates enormously over time. There are two technical terms that help with a discussion of this: conjunction and opposition. While amateur astronomers will be familiar with these words, outsiders should note that the way they are used in this context is counterintuitive in the extreme.

Naively you might think that 'conjunction' refers to the point in their orbits when Earth and Mars are closest to each other, on the same side of the Sun, while 'opposition' means they are as far apart as possible, on opposite sides of the Sun. Unfortunately, it's the other way around. This may sound crazy, but there's an obscure kind of logic to it. The terms go back to ancient times, when everyone thought the Sun and all the planets moved in circles around a stationary Earth. In this model, 'conjunction' means that Mars and the Sun are close to each other on the same side of the Earth, while 'opposition' means they are on opposite sides of the Earth.

Opposition is the best time for observing Mars, because it's closer to Earth and thus appears larger and brighter than usual. Oppositions occur roughly every 25 or 26 months. Even at opposition, however, the distance to Mars can vary considerably depending on whether it's near perihelion or aphelion. The very best observing opportunities occur at perihelion opposition, when Mars is just 0.38 AU from Earth, or approximately 56 million kilometres. Oppositions of this type only come round once every 15 to 17 years – the best-case scenario from an observer's point of view. At the opposite extreme, the worst-case scenario is aphelion conjunction, when Mars is a very distant 2.67 AU from Earth – 400 million kilometres.

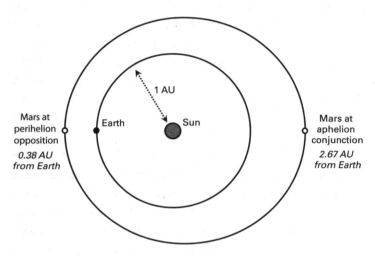

Mars at perihelion opposition
0.38 AU from Earth

Earth

1 AU

Sun

Mars at aphelion conjunction
2.67 AU from Earth

The distance between Earth and Mars varies between two extremes as the planets orbit around the Sun.

It's only since the invention of telescopes that people have been able to see Mars clearly, even at close opposition. Before that it was just another small point of light in the sky, not too different from a bright star or one of the other planets. Even in those days, however, there was something special about Mars. It has a distinctly reddish colour, and the enormous fluctuations in its brightness – between conjunction and opposition, and between perihelion and aphelion – were a baffling mystery to cultures that believed the Earth was the centre of the universe, and that everything in the heavens ought to be perfect and unchangeable. To the ancients, Mars was a symbol of violence and conflict – even our modern name for the planet comes from the Roman god of war – and at one time a bright perihelion opposition was considered a portent of bloodshed to come.

The invention of the telescope in the 17th century did nothing to diminish the fascination with Mars. In fact, it made the Red Planet look even more intriguing, particularly when compared to its Solar System neighbours. Telescopes revealed Mercury to be a small, rocky world like the Moon. Venus, although similar in size to the Earth, is shrouded in thick clouds suggestive of a crushingly dense – and hellishly hot – atmosphere. The outer planets Jupiter and Saturn are huge balls of gas – as are Uranus and Neptune, which are so far away they weren't even discovered until the advent of the telescope. From a human point of view, none of those planets could be described as inviting. But Mars is different.

Viewed through a telescope, Mars resembles a smaller-scale version of our own planet, with a radius of 3,390 km compared to Earth's 6,370 km. Because it's further from the Sun, it receives less of its light; at its brightest, Martian sunlight is about half the intensity of that on Earth. But the length of a Martian 'day' is almost the same as one on Earth: Mars rotates on its axis once every 24 hours and 39 minutes. To avoid confusion with 24-hour Earth days, astronomers call the Martian day a 'sol'.

Topographically, Mars resembles a dryer version of Earth. Its surface features are only hazily visible through a terrestrial telescope, but they give a desert-like impression. It's clear, too, that Mars has an atmosphere – though thinner than that of Earth. The actual density and composition of the Martian atmosphere remained a mystery until the first space probes visited the planet. But even through Earthbound telescopes, astronomers could see that Mars has its own weather – from the occasional wispy cloud to huge dust storms. Like Earth, Mars has ice caps at the north and south poles – although once again, a proper understanding of these had to wait until the space age.

Another Earth-like feature that can be observed on Mars is its regular seasonal cycle. The Earth's seasons come about because its axis of rotation (once every 24 hours) is tilted at 23 degrees relative to its orbit round the Sun (which takes 365 days). Around June, the northern hemisphere is tilted towards the Sun, producing summer in the north and winter in the south. In December the situation is reversed, with winter in the north and summer in the south.

Mars has an axial tilt of 25 degrees, close to that of Earth, so it exhibits a similar cycle of seasons. They're roughly twice as long as Earth seasons, because the Martian year is longer – 687 Earth days, or 669 Martian sols. There's another difference too, caused by the greater ellipticity of Mars's orbit. Perihelion – the closest approach to the Sun – occurs during southern summer and northern winter, while aphelion occurs during southern winter and northern summer. This means that seasonal variations are more pronounced in the southern hemisphere, which tends to have hotter summers and colder winters than the northern hemisphere.

Mars's seasonal variations can be seen from Earth with the aid of a telescope. They're reflected in the expansion and contraction of the ice caps – particularly the southern one – and in changes in the colouring of surface features. In particular, dark areas can be seen to spread out during summer and shrink back in winter. When astronomers first observed this phenomenon in the 19th century, some

Comparison of key data for Earth and Mars

	Earth	Mars
Radius (km)	6,370	3,390
Surface gravity (g)	1	0.4
Solar irradiance (watts per square metre)	1,360	590
Atmospheric pressure (kilopascals)	101	0.6
Perihelion distance to Sun (AU)	0.98	1.38
Aphelion distance to Sun (AU)	1.02	1.67
Length of year (days)	365	687
Length of day (hours)	24	24.6
Axial tilt (degrees)	23	25

of them wondered if they were seeing seasonal changes in native Martian vegetation. This led to one of the most intriguing speculations of the telescopic age: is there life on Mars?

Another Earth?

The year 1877 saw a perihelion opposition of Mars, when the Red Planet was just 56 million kilometres from Earth. This rare opportunity was seized on by an Italian astronomer named Giovanni Schiaparelli, who made detailed observations of the Martian surface. Like all astronomers of the time, he recorded his findings by drawing what he saw through the eyepiece of his telescope. Photography was still in its infancy, and it would be well into the 20th century before cameras were considered reliable enough to be used for professional astronomical work.

Among other surface features, Schiaparelli drew a complex network of straight lines, which he referred to by the Italian word *canali*. Translated into English, this can either mean 'channels' – implying a natural phenomenon – or 'canals', implying an artificial one. Schiaparelli was probably using the word in the first sense, but English-language accounts of his work preferred the more evocative term 'canals'.

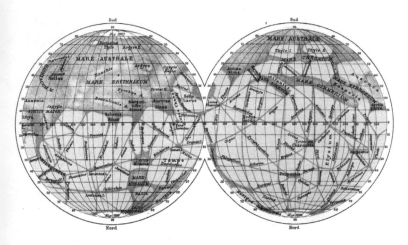

A detailed map of Mars showing its supposed 'canals', produced by Schiaparelli in 1888.

One person who took the idea of Martian canals to heart was the French astronomer Flammarion, who – as described at the start of this chapter – already had a fanciful notion of the Red Planet. Even more ardent was the American Percival Lowell, who made detailed maps of the canals over a period of years, including during the perihelion opposition of 1892. Lowell was convinced the Martian canals were artificial constructions – the work of intelligent creatures who were battling to bring water from the seasonally melting ice caps to the increasingly arid Martian desert. Lowell promoted this view in a series of popular works, starting in 1895 with a book simply titled *Mars*.

The idea of Martian canals was always controversial – not just because it implied the existence of intelligent creatures on Mars, but also because only a handful of people claimed to be able to see them. Many astronomers, looking through telescopes just as good as Schiaparelli's and Lowell's, couldn't see any linear features at all. The matter came to a head at the next perihelion opposition in 1909, when the French astronomer Eugène Antoniadi made the most detailed drawings of Mars to date. There were no canals in Antoniadi's drawings, and he suggested they were nothing but an optical illusion – aided no doubt by wishful thinking:

> The more or less rectilinear, single or double canals of Schiaparelli do not exist as canals or as geometrical patterns; but they have a basis of reality, because on the sites of each of them the surface of the planet shows an irregular streak, or else a broken, greyish border.

That should have been the end of it – and for most professional astronomers, it was. But for the public at large, Martian canals retained a fascination which persisted until space probes established their non-existence beyond any doubt. The canals' popular appeal owed a lot to Lowell's conception of highly intelligent creatures struggling to survive on an increasingly barren planet. The suggestion was that Martian civilisation was older and more advanced than that on Earth. As Lowell himself wrote in his 1895 book:

Quite possibly, such Martian folk are possessed of inventions of which we have not dreamed, and with them electrophones and kinetoscopes are things of a bygone past, preserved with veneration in museums as relics of the clumsy contrivances of the simple childhood of the race. Certainly what we see hints at the existence of beings who are in advance of, not behind us, in the journey of life.

Just two years after Lowell wrote these words, a new serial by H.G. Wells made its appearance in *Pearson's Magazine*. It was called *The War of the Worlds*, and its very first pages portray a vision of Mars almost identical to Lowell's – a dying world populated by super-intelligent Martians. But Wells goes one logical step further than Lowell. If the Martians have advanced technology, and if their planet is dying, wouldn't they start looking for a new one to conquer? Wells's Martians set their greedy sights on Earth, in science fiction's first – but far from last – alien invasion story. The Martians portrayed in *The War of the Worlds* are hideous, tentacled, bug-eyed monsters – a striking, and disturbingly xenophobic, image that would come to dominate mass-market science fiction for decades to come.

A more positive – if less credible – picture of Mars can be found in another magazine serial 15 years after *The War of the Worlds*. Edgar Rice Burroughs's novel *A Princess of Mars* began to appear in *All-Story Magazine* in February 1912. It's the first in what turned out to be a long series of novels, like the same author's better known stories about the jungle hero Tarzan. The Tarzan books may be far-fetched, but the Barsoom novels put them in the shade. 'Barsoom' is the Martians' own name for Mars, as the hero of the first book, John Carter, discovers when he travels there. He doesn't need a spaceship to do that – he simply finds himself teleported to the Red Planet by mystical means. On arrival, he discovers the air is thin but breathable, and Martian topography is much like the American south-west.

During his adventures on Mars, Carter encounters numerous intelligent species, all of them humanoid (unlike Wells's tentacled monsters) – and some virtually indistinguishable from *Homo sapiens*. So indistinguishable, in fact, that Carter eventually marries the eponymous 'Princess of Mars'. Preposterous as it is on the face of it, Burroughs's vision of Mars has one significant feature in common with those of Lowell and Wells – he portrays Mars as a dying world that has seen better days. As the Princess explains in the first novel: 'Were it not for our labours and the fruits of our scientific operations there would not be enough air or water on Mars to support a single human life'.

Credit for the first 'hard' science-fictional treatment of Mars – i.e., in which the science is as painstakingly accurate as the exigencies of the story allow – is usually given to Stanley G. Weinbaum's short story 'A Martian Odyssey', published in *Wonder Stories* in July 1934. This deals with the first trip to Mars – not by mystical teleportation, as in *A Princess of Mars*, but by nuclear powered rocket – as in *The Martian*. And just like Project Ares in *The Martian*, Weinbaum's rocket is called Ares, after the Greek god of war (and counterpart to the Roman god Mars). For modern readers, Weinbaum's description of Mars may come across as full of errors, but it was perfectly respectable in terms of the scientific knowledge of the time. The planet is portrayed as a cold, desert-like place with thin air, which is breathable after special training – not unlike the highest mountains on Earth.

The hero of 'A Martian Odyssey' meets a variety of life forms, all of which reveal a different approach to alien-creation compared to Weinbaum's predecessors. Where Wells thought 'I want to create Martians that are going to invade Earth', and Burroughs thought 'I want to create a beautiful princess that my hero can fall in love with', Weinbaum thought 'I want to create aliens that might have evolved on the planet Mars as it's depicted in factual science books'.

As a result, Weinbaum's Martians really *are* alien – with no points of similarity to anything that's evolved on Earth, no interest in conquering humanity ... and certainly no desire to mate with the protagonist. Perhaps for this reason, 'A Martian Odyssey' made far less impact on the public than either *The War of the Worlds* or *A Princess of Mars*, and it's barely remembered today. Nevertheless, the story played an important role in inspiring a new generation of hard science fiction writers, including Isaac Asimov and Arthur C. Clarke.

Clarke went on to produce his own fictional account of the Red Planet – a novel called *The Sands of Mars*. It appeared in 1951, when Clarke was serving as chairman of the British Interplanetary Society. As such, he was as well-informed as anyone at the time, both about planetary science and about the practicalities of space travel.

The protagonist of *The Sands of Mars* travels to Mars in a nuclear-powered spaceship called Ares – just like 'A Martian Odyssey' before it and *The Martian* 60 years later. Clarke portrays the Martian atmosphere as 'thinner than above the peak of Everest' – requiring the use of a breathing mask but not a full pressure suit. Clarke's Mars is an almost barren desert, with sparse vegetation and a few simple animals. 'Simple' in science fiction terms, that is – meaning about the level of a grazing herbivore (to a scientist 'simple animal' means something like a tardigrade, which needs a microscope to see it properly).

Arthur C. Clarke did his homework. *The Sands of Mars* is an excellent example of hard science fiction, representing the peak of scientifically accurate portrayals of Mars ... until the first space probes arrived and changed everything. At one point in the novel, Clarke confidently asserts (in italics, no less) that *'There are no mountains on Mars'*. Oh yes, there are.

The real Mars

In 1971, 20 years after *The Sands of Mars*, NASA's Mariner 9 became the first spacecraft to go into orbit around the Red Planet. Over a period of several months, it sent back thousands of photographs which transformed our understanding of Martian topography. In this respect at least, it turned out that science fiction had seriously underestimated Mars. As Arthur C. Clarke wrote a few years later in his non-fiction book *The View from Serendip*: 'It has the most spectacular scenery yet discovered anywhere in the universe'. As for those non-existent Martian mountains, Clarke was happy to admit he'd guessed wrong. He pointed to the example of Olympus Mons – an extinct volcano towering 22 kilometres above the surrounding terrain, about two-and-a-half times the height of Mount Everest. Olympus Mons is one of five mountains discovered by Mariner 9 that are higher than anything Earth can offer. Equally spectacular is a giant rift valley, ten times the length and four times the depth of the Grand Canyon in Arizona. It was christened Valles Marineris, after the spacecraft that discovered it.

In the 40-plus years since Mariner 9, around a dozen other robot explorers – orbiters, landers and rovers – have visited the Red Planet. Equipped with sophisticated scientific instruments in addition to cameras, they've sent back vast quantities of ever more detailed data. As a result, we know a lot more about Mars than we used to.

The surface of Mars is mainly composed of volcanic basalt. That's a rock that's found on Earth, too, but Martian basalt is particularly rich in iron oxide – or rust, to use its non-scientific name. It's what gives the 'Red' Planet its distinctive orange-brown hue. Seen by robot landers on the surface, the Martian sky is a reddish colour too, caused by fine dust particles suspended in the atmosphere. Dust is one thing Mars has plenty of – most of the surface is covered with it. Being a very dry planet, the dust gets blown around a lot; huge dust storms are the most distinctive and dramatic features of Martian weather.

Because Mars is smaller than Earth, its surface gravity is only about 0.4 g ('1 g' is the pull of gravity at the Earth's surface). This is one of the reasons why Martian mountains are so big – there is less gravity to flatten them out. Mars's gravity does, however, control more natural satellites than the Earth, with two rocky moons orbiting the planet rather than one. But the Martian moons, called Phobos and Deimos, are nothing like the Earth's Moon. They're less than a hundredth its size, and closer in shape to a potato than a sphere. They look like asteroids – the small rocky bodies mainly found in a broad belt between Mars and Jupiter – and that may be exactly what they used to be, before they were captured and pulled in by Martian gravity. The larger of the two, Phobos, is only about 25 km across, but it whizzes around Mars in a much lower orbit than our own Moon, just 6,000 km or so from the surface. This makes it appear quite large in the Martian sky – up to a third the size of the full Moon seen from Earth.

Viewed through the electronic eyes of one of NASA's rovers, Mars looks rather Earth-like – or at least, like some of the drier and more barren parts of Earth. Cameras only tell part of the story, though. Other instruments and sensors make it clear that Mars is an inhospitable place indeed. The atmospheric pressure at the surface is only about 0.6 per cent of that on our own planet. That's comparable to the air on Earth at an altitude of 35 km. The Martian atmosphere differs in its composition, too. It's mainly carbon dioxide, with only trace amounts of oxygen and water vapour. This sparse atmosphere doesn't just mean there's nothing to breathe; it makes it impossibly cold at night, too. The daytime temperature can be as high as 20 degrees Celsius – which doesn't sound too bad – but during the hours of darkness it can plunge to minus 140 degrees. That's almost as cold as it would be without any atmosphere at all.

When Percival Lowell was fantasising about Martian canals at the end of the 19th century, he imagined they were designed to channel seasonally melting ice from the north and south poles to the more arid equatorial regions. This was based on the observation, which can easily be made from Earth, that the Martian polar caps expand during winter and shrink during summer. And what else could the poles be made of but water ice?

It turns out that Lowell was right and wrong at the same time. When the first space probes measured the Martian atmosphere, it became clear that the seasonally changing polar caps weren't made of water ice but carbon dioxide, which freezes out of the atmosphere during winter and evaporates again in summer. However, subsequent radar observations revealed that there really *is* water ice at the poles – and a lot of it. It's deeper down, and not as easily visible as the frozen carbon dioxide, but there's much more of it – several million cubic kilometres in total.

Mars's low surface temperature means that subsurface ice isn't restricted to the poles – it's found at lower latitudes too. If all the ice found on Mars was spread out evenly, it would cover the whole planet to a depth of several metres. That's good news for future human explorers, who will be able to obtain as much water as they want simply by 'quarrying' ice. There's even evidence that the subsurface ice can sometimes melt, when the weather gets warm enough. In 2010, NASA's Mars Reconnaissance Orbiter found a number of narrow, dark markings on steep, well-illuminated slopes, which were dubbed 'recurring slope lineae'. These typically occur in places where the surface temperature is above minus 20 degrees Celsius, and appear to be caused by a liquid flowing downhill. Of the various explanations that have been put forward, the likeliest is that they're caused by extremely briny water, with salt concentrations much higher than seawater on Earth. This has the effect of lowering the freezing temperature – the same principle behind the gritting of roads with rock salt in winter.

So there's water on Mars, but it's elusive. That's not always been the case, though. Martian geology shows plenty of evidence for flowing water in the distant past. Valles Marineris, for example, although it was originally formed by geological faulting, was subsequently shaped by water erosion. Similar processes can be seen all over the planet, in the form of dried-up riverbeds and lakebeds. Other circumstantial evidence includes rounded pebbles, which on

Earth are usually formed by the action of running water. The fact that large bodies of water once existed on the surface implies that the Martian atmosphere must have been much thicker than it is today – thick enough to produce a temperate climate.

So those 20th-century science fiction writers were right after all – the Red Planet used to be a much warmer, wetter place than it is now. But they didn't get it quite right. Science fiction generally portrayed 'wet Mars' as having existed a few centuries ago, with the last struggling descendants of those times still around today. That's not the case, though. By comparing water-formed features with overlying impact craters of known age, scientists estimate that Mars's wet period may have ended as long as three billion years ago.

Three billion years ago, there was life on Earth – and it had already been evolving for almost a billion years. Yet it was still extremely simple life, consisting of microscopically small, single-celled organisms. It was only later – perhaps a billion and a half years later – that more complex, multicellular creatures appeared, and even then they were still microscopic. It's conceivable that life arose on Mars around the same time as on Earth, or even slightly earlier. There's little doubt the right sort of conditions existed. But it's very doubtful that Martian life progressed beyond microscopic forms. Creatures large enough to interest a science fiction writer – things like fish and land animals – only made their appearance on Earth in the last half-billion years. By that time, Mars had been a virtually airless and waterless wasteland for over two billion years.

Even if the surface of Mars did harbour simple life forms at some point in the past, very few scientists hold any hope of finding life there now. Those seasonally spreading patches of 'vegetation', which so excited 19th-century astronomers like Percival Lowell, are no such thing. Unlike Schiaparelli's canals, the dark patches are real – not an optical illusion – but they're simply the result of ever-shifting patterns of wind-blown Martian dust.

Nevertheless, Mars may not be completely dead. It's possible there's liquid water far below the surface, heated – as in some deep caves and fissures on Earth – by the planet's internal energy. Where there's water, there may be life. That's the case in Earth's deep biosphere, which extends kilometres below the surface and harbours 'extremophiles' that have learned to thrive in such environments. For the most part these are simple, single-celled organisms like bacteria, but more complex worm-like creatures have also been found. Many scientists believe similar extremophiles could lurk deep down in the rocky crust of Mars, too.

No Martian canals, no bug-eyed monsters, no beautiful princesses ... maybe a few bacteria or worms if we're lucky. Mars just isn't as interesting as it used to be. At least, not for people who rate planets solely by their likelihood of harbouring intelligent life. Amazingly, there are still a few individuals who – chiefly on the internet – continue to promote the idea that Mars was home to a highly advanced civilisation in its archaeological past. They scour photographs from NASA's orbiters and rovers looking for evidence to support their beliefs – relying, just as Percival Lowell did, on optical illusions and wishful thinking. The best-known example of this is the 'Face on Mars', a large rocky outcrop that – in images taken in the 1970s – looked for all the world like a stylised human face. More recent photographs, taken with higher resolution cameras, dispel this illusion for all but the most hardened believers. An article on the NASA website entitled

'Unmasking the Face on Mars' makes it clear the object in question is a natural formation like the buttes and mesas of the American West. After dismissing any possibility of an archaeological relic, the same article notes that 'defenders of the NASA budget wish there *was* an ancient civilization on Mars'.

This is an important point. People who want to send spacecraft to Mars – whether robot probes or human missions – have their own good reasons for doing so. But to get the necessary funds and commitment, they need to spark the interest of politicians, the media and the public. Nothing would do that better than discovering that Mars was once home to an intelligent species like our own. Even a lower life form – such as the one portrayed by Arthur C. Clarke in *The Sands of Mars* – would do the trick if it was still chomping its way through the Martian vegetation today. But we don't have any of that – not even the faintest, tantalising hint.

So why go to Mars? Actually there's a very good reason. Mars may not be inhabited – yet – but there's no doubt it's *habitable*, with a little effort. As seen in the table on page 7, it's really quite Earth-like. It's got plenty of water, even if this needs to be extracted from subsurface ice. It's got sunlight, which can be converted to electrical power using solar panels. As in *The Martian*, humans would have to live inside pressurised habitats – and again as in *The Martian*, it's possible to extract the necessary oxygen from atmospheric carbon dioxide. And *The Martian* was right about another thing, too: it would be possible to grow Earth food in Martian soil. When NASA's Phoenix probe landed in May 2008, the first soil sample it took turned out to be surprisingly plant-friendly. As one of the mission scientists put it at the time: 'It is the type of soil you would probably have in your back yard, you know, alkaline. You might be able to grow asparagus in it really well.'

From the point of view of human life, Mars may not be the most hospitable place in the Solar System – that's Earth – but it comes a not-too-distant second. Which brings us to the next question – how do we get there?

HOW TO GET TO MARS 2

Rocket science

At the start of the 20th century, the idea that a rocket could be used to travel to another planet was very much the stuff of science fiction. The American engineer Robert Goddard was among the few people who took the subject seriously. His writings on extraterrestrial rocketry tended to meet with ridicule – not because of any supposed problems with his engineering designs, but for a much more fundamental reason. According to the popular view at the time, the use of rockets beyond the Earth's atmosphere was a scientific impossibility. The most famous version of this argument appeared in a *New York Times* editorial in January 1920:

> After the rocket quits our air and really starts on its longer journey, its flight would be neither accelerated nor maintained by the explosion of the charges it then might have left. To claim that it would be is to deny a fundamental law of dynamics. [...] That Professor Goddard, with his chair

in Clark College and the countenancing of the Smithsonian Institution, does not know the relation of action to reaction, and of the need to have something better than a vacuum against which to react – to say that would be absurd. Of course he only seems to lack the knowledge ladled out daily in high schools.

The *New York Times* was wrong – and 49 years later, with Apollo 11 en route to the Moon, the newspaper issued a belated correction:

> Further investigation and experimentation have confirmed the findings of Isaac Newton in the 17th Century and it is now definitely established that a rocket can function in a vacuum as well as in an atmosphere. The *Times* regrets the error.

The reference to Newton is crucial. To a good approximation, everything you need to know about the science of space travel can be found in a book he wrote in 1687 – *Philosophiae Naturalis Principia Mathematica*, or 'Mathematical Principles of Natural Philosophy'. As well as his famous law of gravity, the book contains Newton's laws of motion. In essence, all three of these laws are statements of the same basic principle – the conservation of momentum.

The momentum of an object is defined as its mass multiplied by its velocity. For most objects mass is more or less constant, so momentum is essentially proportional to velocity. You know what speed your car is doing, so who cares about its momentum? Unfortunately, things are more complicated when it comes to rockets. They're constantly *losing mass* – that's the very principle on which they work. So we really do have to talk about a rocket's momentum – its (variable) speed times its (variable) mass.

The net increase in forward momentum resulting from a rocket burn is called the 'impulse', which turns out to be equal to the force of the rocket's thrust multiplied by the duration of the burn. But why does burning fuel in a rocket engine end up pushing the rocket along? It's not the actual combustion that produces the thrust, but the exhaust – the waste gases that are pushed out of the back of the rocket (that's why it's continually losing mass). This is where conservation of momentum comes in – the momentum of the exhaust flying out backwards has to be balanced by an increased forward momentum of the rocket itself.

It's possible to use Newton's laws of motion to derive a simple formula called 'the rocket equation'. This was done as long ago as the 19th century by the Russian engineer Konstantin Tsiolkovsky, one of the great pioneers of rocketry. Tsiolkovsky's formula relates the change in speed – usually referred to as 'delta-v' – to the speed of the rocket's exhaust and the ratio of its initial to final mass.

Science fiction fans may recognise 'delta-v' as a buzzword authors use when they want to indicate that a character is a rocket scientist. But the authors are right: rocket scientists do talk about delta-v – even in real life. Down on terra firma, car manufacturers obsess about acceleration – for example, the number of seconds taken to get from zero to 100 km/h. But in space, the important thing is the *difference* between the final and initial speeds (100 km/h in that example). That's the delta-v. It doesn't matter if it takes a few seconds or several hours, as long as the desired delta-v is achieved.

The reason delta-v is so important is that space travel is all about orbits, and an orbit is ultimately characterised by its velocity. To put a spacecraft into a desired orbit, or to match orbits with another object, it's not enough to get to a point in space – you must be travelling at exactly the right speed, in the right direction, when you reach it. To understand this properly we need to go back to Isaac Newton, and the most famous of all his laws – the law of gravity.

Everyone knows what gravity is – it's a force of attraction that acts towards the centre of a planet, and gets weaker as the distance to the planet increases. Newton's law gives the exact formula – the force of gravity varies in proportion to the mass of the planet and the inverse square of the distance from its centre. From everyday experience, we know things tend to fall to Earth. That would be true out in space too, if the object we're talking about is initially stationary or moving very slowly. But if it's moving fast – as most spacecraft are – then conservation of momentum tends to keep it moving in a straight line. That's Newton's first law of motion, also known as the principle of inertia.

The net effect of gravity and inertia on a fast-moving spacecraft is to *bend* its trajectory. The circular orbit of a satellite is the result of a fine balancing act. If inertia had its own way, the satellite would simply fly off at a tangent. On the other hand – and even more obviously – if gravity was the only consideration the satellite would plummet down to Earth. In reality, both factors are at play – gravity and inertia – and the result is an orbit, just like the orbit of a planet around the Sun.

The word 'orbit' usually conjures up the image of a spacecraft following a circular trajectory around the Earth, but that's not the only possibility. Intersecting any point on that circular orbit, there are an infinite number of other orbits with different speeds in different directions. A spacecraft going significantly faster than the circular orbit will loop out on an elongated path, taking it to a much higher altitude on the opposite side of the Earth. If it's going fast enough – faster than Earth's 'escape velocity' – it won't be in orbit around the Earth at all. The spacecraft will still follow an orbit, but it's an orbit around the Sun instead of the Earth. Depending on its exact velocity, this orbit could take the spacecraft to Venus, or Jupiter, or Neptune ... or Mars. That's why delta-v is so important.

The science fiction writer Jerry Pournelle has described Earth orbit as 'halfway to anywhere'. On the face of it, that's a nonsensical claim. Orbiting satellites are just a few hundred kilometres above the Earth's surface, while Mars is some 60 million kilometres away at its closest. Jupiter is ten times further than that, and the distance to Neptune is over four billion kilometres. But Pournelle wasn't talking about kilometres. Like any good rocket scientist, he was thinking in terms of delta-v.

In the absence of atmospheric drag, getting a spacecraft from the Earth's surface into a circular orbit would require a delta-v of 7.8 km/s. In practice, to overcome air resistance, a somewhat higher figure, around 10 km/s, is needed. That's an easy number to remember – and it's worth remembering too, because it's the minimum effort you need for that first step up into Earth orbit. But does that really put you 'halfway to anywhere'?

Yes it does – if you're thinking in terms of delta-v rather than kilometres. An additional delta-v of just 3.6 km/s will get you out to the orbit of Mars, 6.3 km/s to the orbit of Jupiter and 8.2 km/s to the orbit of Neptune. Back in 1977 NASA launched the space probe Voyager 1 with a total delta-v of 18.3 km/s relative to the Earth's surface – still less than twice what's needed to reach Earth orbit – and it has now left the Solar System and is on its way to the stars.

Action and reaction

A rocket works on much the same principle as a jet aircraft. The jet burns fuel inside a combustion chamber to create an extremely hot gas, which is then forced out through an exhaust nozzle to give the aircraft a forward thrust. But the jet can use a couple of tricks that a rocket can't. From a technical point of view, 'burning' is a chemical reaction between the fuel and oxygen – and the aircraft can suck in all the oxygen it needs from the atmosphere. On top of that, a jet engine can supplement the hot exhaust with any quantity of cool air – again sucked in from the atmosphere and blown out through the back of the engine – to boost the amount of thrust it produces. Rockets just can't do that. They have to take everything along with them – fuel, oxygen and anything else that's flung out of the back to generate thrust.

This gives rise to one of a rocket scientist's biggest headaches. A launch vehicle needs enough power, not just to lift the payload off the ground, but to lift its own mass too. The more powerful the rocket, the more fuel, oxygen and propellant it needs – and the more of those it carries, the larger and heavier the rocket. For a jet, the fraction of the vehicle's total mass which represents useful payload may be close to 50 per cent. For a space launcher, the 'payload fraction' is more like 3 or 4 per cent. Only a small part of the huge rocket that lifts off from the launch pad will ever acquire enough delta-v to enter orbit.

Back in 1952, five years before the first artificial satellite was sent into orbit, Isaac Asimov wrote a short story called 'The Martian Way' which anticipated the problem very clearly. It all stems from Newton's third law of motion – the law of equal and opposite action and reaction (another consequence of the conservation of momentum). As one of the characters explains:

> Now imagine a spaceship that weighs a hundred thousand tons lifting off Earth. To do that, something else must move downward. Since a spaceship is extremely heavy, a great deal of material must be moved downwards. So much material, in fact, that there is no place to keep it all aboard ship. A special compartment must be built behind the ship to hold it. [...] But now the total weight of the ship is much greater. You will need still more propulsion and still more. [...] When the material inside the biggest shell is used up, the shell is detached. It's thrown away too. [...] Then the second one goes, and then, if the trip is a long one, the last is ejected.

Words like 'compartment' and 'shell' may differ from the terminology now used by rocket engineers, but what Asimov was getting at is the essential need for rocket staging. This is a fundamental reality of rocket science, and one that Tsiolkovsky was aware of back in the 19th century. However, staging is such a clunky and inelegant process that most science fiction writers – before and after Asimov – have tended to ignore it. Unfortunately, their preferred option of a 'single-stage-to-orbit' is still a dream of the future. To this day, any practical launcher has to be a multi-stage vehicle, made up of two or more rockets stacked on top of each other. That way – just as Asimov described – the dead weight of a spent stage can be jettisoned as soon as it runs out of fuel.

All current space launchers use chemical rockets, which get their energy by burning fuel and oxygen just like a jet airliner. The fuel may be a type of kerosene similar to standard aviation fuel, or for higher performance it may be pure hydrogen in highly compressed liquid form. Since obtaining oxygen from the atmosphere isn't an option, these rockets also have to take their own oxygen along – again compressed into liquid form. One of the advantages of a chemical-type rocket is that it kills two birds with one stone. Firstly, the combustion of fuel and oxygen generates a large amount of energy in the form of heat. Secondly, expelling the superheated combustion products through a narrow exhaust nozzle produces the necessary thrust to push the rocket aloft.

There are limits to what can be achieved with chemicals, however. The effectiveness of a rocket system can be characterised by three factors. Two are easy: the rocket's thrust, usually measured in newtons (a newton is the force needed to raise the speed of a 1 kg object by 1 metre per second every second), and the time for which this thrust can be applied. The third factor is more obscure, but just as important – the speed at which exhaust particles are expelled. The higher this exhaust speed, the bigger the push imparted to the rocket by a given quantity of propellant.

That word, 'propellant', is often used interchangeably with 'fuel'. In the case of a chemical rocket, there's no problem with this. The propellant really is the fuel – or more pedantically, it's the hot exhaust gas created by burning the fuel and oxygen. This type of rocket has a low exhaust velocity, just a few kilometres per second, but it makes up for it by burning vast quantities of fuel very quickly. In this way, a large chemical rocket can generate an enormous thrust, of the order of several million newtons – but only for a few minutes before all the fuel is used up.

All rockets need fuel, and all rockets need propellant. But the fuel and propellant don't have to be the same thing. The propellant is what gives the rocket a push through space, and the fuel is what gives the propellant its energy. Chemical combustion is a time-honoured way to produce energy, but it's far from the most effective. Nuclear fission reactions, for example, can produce around a million times more energy per kilogram than a chemical reaction. Just a few kilograms of uranium can keep a 15,000-tonne nuclear submarine going for a year. So why not use a nuclear reactor to power a rocket?

The problem is, left to itself a nuclear reactor produces energy but it doesn't produce thrust. There are no naturally produced exhaust gases to serve as a propellant. A nuclear-powered rocket needs a second component, in addition to the energy-producing reactor – something that generates a fast-moving stream of propellant to push the rocket along. The simplest way to do this is to use the reactor to heat a gas, which then functions as a propellant like the exhaust from a chemical reaction. Such an arrangement is called a 'nuclear thermal rocket', and it highlights one of the great ironies of rocket science. No matter how compact and efficient its energy source, a rocket *still* has to carry a load of propellant along with it. Miraculous science fiction space drives that ignore this principle are just that – science fiction.*

---

* Or are they? Over the last few years, publicity has been given to so-called 'reactionless drives' – the EmDrive being the best-known example – which really do appear to circumvent the known laws of physics. The jury is still out, although most scientists remain deeply sceptical about such claims. More will be said on this subject in the final chapter.

There is nothing science-fictional about a nuclear thermal rocket, though. A design of this type was considered by NASA in the 1960s, and featured in early plans for a mission to Mars. Called NERVA (Nuclear Engine for Rocket Vehicle Application), it was a potential replacement for the third stage of the giant Saturn V launcher. The main purpose of this stage is not to get into Earth orbit – that's the job of the first and second stages – but to generate the extra delta-v needed to break free of Earth's gravity and shift on to an interplanetary trajectory. NERVA would have achieved this more efficiently than the standard, chemically powered third stage, with twice its exhaust velocity. Both engines produced a similar thrust, in the region of a million newtons, but NERVA could keep this up for three times as long – half an hour – before the propellant was used up.

The result was a significantly higher delta-v ... or it would have been, if NERVA had ever made it into space. As it turned out, the project was abruptly cancelled by Congress in 1972 – for political reasons rather than any technical shortcomings. One person who campaigned against the project's cancellation was science fiction author Jerry Pournelle, mentioned earlier in this chapter. He expressed his bitterness at Congress's decision in the following way:

> More money is annually spent on lipsticks in New York state than NERVA was costing, and any medium-sized state has more annual sales of liquor than NERVA cost over its lifetime; but it's nice that they don't waste the taxpayers' money with frivolities like space.

NERVA had one thing in common with conventional chemical rockets – it used heat energy to accelerate the propellant to the required exhaust velocity. Another option is to use electromagnetic energy instead. In this case the propellant can't be a normal gas, because gases are electrically neutral and so don't feel the effect of electric or magnetic fields. Instead, an 'electromagnetic rocket' uses a special kind of gas called a plasma, made up of electrically charged ions rather than neutral atoms. An ion-based engine needs not one but two clever pieces of technology – the first to convert an ordinary gas into a plasma, and the second to accelerate the plasma to a suitable exhaust speed.

These things can be achieved in different ways. One of the simplest designs is the ion thruster, as used on several NASA spacecraft including the Dawn mission launched in 2007. Dawn travelled out to the asteroid belt between Mars and Jupiter – a region littered with over a million space rocks. Some of them are bigger than others – Dawn's first stop, Vesta, is the second largest at 500 km across. Arriving there in 2011, Dawn spent a year in orbit, taking photographs and making scientific measurements, before setting off again. Its next target was Ceres – the largest object in the asteroid belt, where it entered orbit in 2015. Almost twice the size of Vesta, Ceres is more spherical in shape, looking less like an oversized boulder and more like a miniature planet. In fact Ceres is officially classed as a 'dwarf planet', like Pluto, rather than an asteroid.

Dawn's ambitious itinerary would have been unthinkable if it had been equipped with conventional rocket thrusters, but the high-tech ion drive made it all possible. Dawn's basic power source is not chemical fuel but the Sun – courtesy of its two 8.3 by 2.3 metre solar arrays. Electrical power is used to ionise the xenon gas propellant, which is then accelerated to high speed using a strong electric field. The resulting

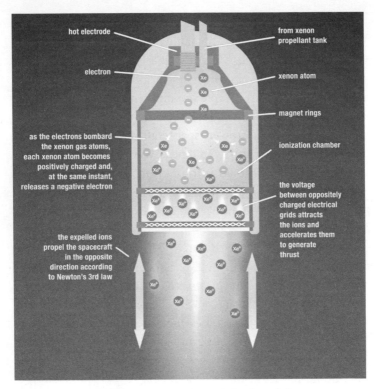

hot electrode

electron

as the electrons bombard the xenon gas atoms, each xenon atom becomes positively charged and, at the same instant, releases a negative electron

the expelled ions propel the spacecraft in the opposite direction according to Newton's 3rd law

from xenon propellant tank

xenon atom

magnet rings

ionization chamber

the voltage between oppositely charged electrical grids attracts the ions and accelerates them to generate thrust

**Schematic illustration of the ion drive
used by the Dawn spacecraft.**

(NASA image*)

exhaust velocity is over 30 km/s – ten times that of a chemical rocket. This doesn't produce a huge burst of acceleration, though, because the thrust of the ion engine is tiny – just 90 millinewtons. But this thrust can be kept up for much longer periods of time, so it's still capable of producing the necessary delta-v in the end.

---

* The full graphic can be found at http://dawn.jpl.nasa.gov/ multimedia/images/DawnIonPropulsionInfographic.jpg

Of course, the tortoise-versus-hare ion thruster used on Dawn would be completely unsuitable for a human mission. As a robot, Dawn has inexhaustible patience, and doesn't have to worry about carrying enough food, air and water for a decade-long mission. People are different. Fortunately, there are other variations on the ion thruster theme that might be more suitable for a crewed spacecraft. One such design, in fictional guise, is featured in Andy Weir's novel *The Martian*. Although he doesn't go into any great detail, his engine appears to be based on the VASIMR concept – the Variable Specific Impulse Magnetoplasma Rocket. This is a real design, which has been championed for many years by former NASA astronaut Franklin Chang Díaz. VASIMR uses radio waves to create a hot plasma, followed by a magnetic rather than electric field to accelerate the plasma to exhaust velocity. Although it has never flown in space, VASIMR engines have been tested on the ground. They fall somewhere between the extremes of high-thrust conventional rockets and low-thrust ion drives, producing a thrust of a few thousand newtons which can be kept up for months or even years.

Where does a VASIMR engine get its energy from? In *The Martian*, the answer is a nuclear reactor. This is another way of using nuclear power in space – 'nuclear electric propulsion', as opposed to the nuclear thermal propulsion of NERVA. Nuclear electric propulsion is one of the options NASA is considering in its Mars plans – but only an outlier, because of the social and political negativity associated with anything nuclear.

In fact, a few small nuclear reactors have already been sent into space – mainly by Russia – but only to provide electrical energy for satellite systems, not for propulsion. There's potential for confusion here with a much commoner piece of space technology, called a radioisotope thermal generator, or RTG. NASA often uses these as a convenient way to produce electricity, when the more familiar option of solar panels would be impractical. An RTG isn't a nuclear reactor, but it does use a highly radioactive isotope of plutonium, so it raises similar environmental issues. As well as producing radioactivity, an unstable isotope also generates considerable heat. An RTG exploits this natural process by converting heat to electrical energy. NASA has already put three RTGs on the surface of Mars – in the two Viking landers of the 1970s, and more recently in the Curiosity rover. This shows that it's perfectly feasible to launch 'nuclear' systems into space – although in the context of a human mission to Mars, most people will be happier if the idea remains in the background as Plan B.

The long way round

An orbiting spacecraft is still bound to Earth by the force of gravity. To pull free of that force, it needs to achieve escape velocity – roughly another 3 km/s of delta-v on top of orbital velocity. Only then can it turn its attention to Mars.

And what's the quickest way to Mars? How about this. Wait until the Red Planet is approaching opposition, which it does every couple of years. At this point it could be as little as 60 million km away. If we point the spacecraft in the right direction and go flat out, we'll be there in no time at all. If we've already reached a comfortable escape velocity of 15 km/s, the journey should take about 4 million seconds, or a few days short of seven weeks. That's pretty quick, by space travel standards.

Unfortunately, things aren't that simple. That escape velocity that you've struggled so hard to achieve is measured *relative to Earth*. And the Earth is already travelling at 30 km/s around the Sun – exactly at right angles to the direction you need to travel to get to Mars. Whether you like it or not, you're going to share that motion, even if you're no longer gravitationally bound to the Earth. To get to Mars by the straight-line route, you must first lose all that 30 km/s. Then when you start to approach the Red Planet, you'll find you're travelling in the wrong direction again. Mars is orbiting the Sun at 24 km/s – at right angles to the direction you're going in – and you need to match speeds with it. Overall, this way of travelling to Mars is going to require an absurdly large delta-v.

If you can't get to Mars by travelling in a straight line, how do you make the trip? Because we're talking rocket science, the answer is typically counterintuitive. The easiest way to get to Mars is to take the long way round.

As long ago as 1935, a German scientist named Walter Hohmann worked out a clever way to get from one planetary orbit to another using the least amount of energy. The route he came up with is now referred to as a 'Hohmann transfer orbit', and most interplanetary space missions use something close to it. To travel from the Earth to Mars, for example, you need to follow an elliptical orbit around the Sun which has its perihelion at Earth (your starting point) and its aphelion at Mars (your destination). As you can see from the diagram on the next page, by the time you get to Mars it's on the opposite side of the Sun from your starting point. That may look like it requires a lot of fuel, but it doesn't. Remember you're always travelling *relative to the Earth*, and the Earth's natural motion – its yearly orbit around the Sun – does most of the hard work for you.

Hohmann's route to Mars is amazingly economical in terms of delta-v, and depending on the exact geometry, it can take as little as eight months to arrive at the orbit of Mars. Of course, you need to do more than just get to Mars's orbit – you need to get to the specific point on that orbit where Mars happens to be at the time. To make sure that happens, you have to get your timing exactly right. Suitable launch windows are relatively rare occurrences – they crop up once every two years and two months, and usually only stay open for a matter of days.

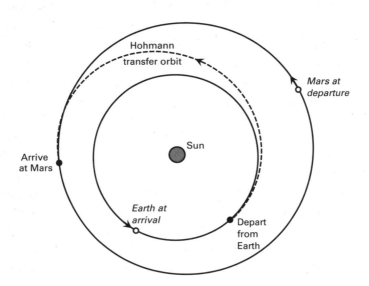

Hohmann transfer orbit to Mars.

For a human mission to Mars, there's another important factor to consider: how to get back to Earth. There are two main options here, which planners refer to as opposition and conjunction class missions. They differ in two important ways: how much energy they use, and how long the astronauts stay on Mars before setting off back to Earth.

The long-stay option uses a standard Hohmann trajectory for both outward and return journeys. A characteristic of these trajectories is that they finish up on the opposite side of the Sun from their starting point – so, using that counterintuitive logic described earlier, this option is referred to as a 'conjunction class' mission. Because the astronauts have to wait for the return Hohmann window, there's no alternative but to remain on Mars – or in orbit around it – for about 16 months. Of the two options, this one has the advantage that it requires the least energy, with a total round trip delta-v (on top of Earth escape velocity) of around 8 km/s. The downside of this approach is that the astronauts must spend a long time in space – almost three years – with all the implications for life support that entails.

The shorter alternative is to use a standard Hohmann route on just one leg of the journey – either outbound or return – and a different trajectory on the other leg. The latter can be approximated as two Hohmann trajectories, one between the orbits of Mars and Venus, and the other between the orbits of Venus and Earth. This route ends up on the same side of the Sun as its starting point, so it's referred to as opposition-class. It may sound like an unnecessarily roundabout route, but it has the advantage that the stay on Mars can be as short as 30 days, resulting in a total time away from Earth of just 18 months or so. This comes at a price, though – the total delta-v may be as much as 12 km/s higher than for a conjunction-class mission.

Although it's not stated explicitly in the novel, the Ares 3 mission portrayed in *The Martian* must use an opposition-class profile. This means that, when the mission is aborted after just six days on Mars (or 18 days in the film version), the astronauts can return straight to Earth. If a conjunction-class mission had been aborted after such a short time, they would have been forced to remain in orbit around Mars for more than a year before they had the opportunity to return.

What if a mission had to abort before it had even arrived at Mars? The nature of the Hohmann orbit means there's no turning back – not even if the spacecraft is only a few weeks out from Earth when the emergency arises. Just like the Apollo 13 astronauts, the crew would have to carry on all the way to their destination, swing around it (Mars in this case, rather than the Moon) and then travel all the way back to Earth. Some people have even proposed this sort of fly-by mission profile as a way of 'doing' the Red Planet, bucket-list style, rather than landing and carrying out any real science.

Most people, however, would say there's no point going all the way to Mars if you're not going to spend at least a little time looking around. And that raises a whole bunch of new problems.

Landing on Mars

Entering orbit around Mars requires about 2 km/s of delta-v, but not all of that needs to come from the spacecraft's engine. If it skims close enough to the planet, some of the necessary delta-v can be obtained from the braking effect of Mars's atmosphere, completely free of charge. Referred to as 'aerocapture', this manoeuvre hasn't been used in practice yet – mainly because it hasn't been needed for the size of orbiters sent to Mars to date. They've only needed a modest amount of fuel to achieve the necessary delta-v using a rocket burn. However, such orbiters often make use of a related technique, called aerobraking, to shift from a high, elliptical orbit to a lower, more circular one.

Once a suitable orbit has been achieved, the next step is the trickiest of all – entry, descent and landing, often abbreviated to EDL. In the case of a human mission, the crew would transfer to a small detachable spacecraft – the Martian equivalent of the Apollo lunar module. But there's an important difference here. The lunar module included both a descent stage and an ascent stage in a single vehicle, which was perfectly adequate to deal with the weak pull of the Moon's gravity. The same trick wouldn't work for Mars, where escape velocity is twice that of the Moon. For that reason, most current plans have a Mars Ascent Vehicle, or MAV, predeployed on the surface before the astronauts arrive. All the descent vehicle needs to do is get them down to the surface. Of course, it has to do that pretty accurately – within walking distance of the MAV and all the other predeployed equipment.

Landing on Mars is more like returning to Earth from space than landing on the Moon. In the latter case, there's absolutely no atmosphere – so the descent can be controlled entirely by a downward-pointing rocket engine. The usual purpose of a rocket is to make something go faster, but in this case the aim is the opposite – to slow the lander down. Rockets used in this way are sometimes referred to as 'retro-rockets'.

A spacecraft attempting to land on Earth in the same way the lunar module landed on the Moon wouldn't make it very far – it would burn up in the upper atmosphere. This isn't because the upper atmosphere is a ferociously hot place, but because the enormous speed of an incoming spacecraft generates a huge amount of heat through aerodynamic compression. A returning spacecraft has to make the aerodynamics work for it instead of against it. Those aerodynamic forces can be used to shed a lot of the spacecraft's speed, by converting it into heat – which must then be dissipated using a carefully designed heat shield. In the case of the Space Shuttle, re-entry was followed by a controlled glide through the atmosphere using the aerodynamic lift of its wings. Smaller space capsules like Soyuz use parachutes to slow them as they fall, with small retro-rockets firing for a second or so before touchdown to soften the impact.

Landing on Mars is even harder. It has an atmosphere – not a very thick one, but thick enough to produce frictional heating. So a rocket-only system of the kind used for lunar landings wouldn't work. On the other hand, the atmosphere is much thinner than Earth's, so bringing a vehicle to a controlled landing using aerodynamics or a parachute is decidedly tricky. Over the last 40 years, robot landers have tried various combinations of atmospheric braking, aerodynamic control, parachutes, retro-rockets … and a couple of new tricks, too. And it's safe to say they haven't always worked.

MARTIAN ROBOTS 3

Fifty years of Mars exploration

On 20 July 1976 – exactly seven years on from the day Apollo 11 touched down on the Moon – NASA's Viking 1 lander separated from its mother craft in orbit around Mars. Over the next three hours it descended through the thin Martian atmosphere, finally settling on to the surface of the Red Planet at 11.53 GMT. For the first time in history, a fully functioning spacecraft had arrived on Mars.

But there were no people on board Viking 1. It was just a robot – albeit an enormously sophisticated one by the standards of the 1970s: 'an incredible *tour de force* of technology, containing forty thousand components in the volume of a biscuit tin', as Arthur C. Clarke put it. The complete Viking spacecraft – orbiter and lander – weighed less than three tonnes, and no human-carrying vehicle could possibly compete with that. In a field where size means money, robots have dominated interplanetary space flight for the last half-century.

The first robotic probe to reach Mars was NASA's Mariner 4. This didn't land or go into orbit – it just flew past the planet, taking a few photographs as it did so. Weighing 260 kg, Mariner 4 was launched in November 1964 on top of a two-stage rocket. The first stage and an initial burn of the second stage were needed to reach Earth orbit, followed by a second burn of the upper stage to shift into Mars transfer orbit. Although the mass of the spacecraft – and the power of the launch rocket – has gradually increased, all subsequent Mars missions have been launched by much the same method.

Sending a robot to Mars is only half the problem. It's also essential to have a good two-way communication link, so the probe can send images and other data back to Earth, and ground controllers can send commands to the spacecraft. By the time Mariner 4 was launched, NASA had set up its Deep Space Network (DSN) for exactly this purpose. In its original form, this consisted of three large radio dishes, roughly equally spaced around the Earth. There was one in Australia, one in South Africa and one at Goldstone in California. This planet-wide array allowed NASA to keep in touch with any interplanetary spacecraft as long as it had a clear line of sight to Earth. The same basic network is still in use today, although with upgraded equipment and one change of venue – the South African station was replaced by a site near Madrid in Spain in the 1970s.

Mariner 4 made its closest approach to Mars on 15 July 1965, at which time it was just under 10,000 km from the Red Planet. Because of the speed at which it was travelling, Mariner 4's big photo opportunity lasted a mere 26 minutes, during which time it snapped just 21 pictures. Sending them back to Earth took considerably longer. The technology of the time only permitted a data rate of 8.3 bits per second, meaning that it took more than eight hours to send a single picture. Allowing for interruptions when the spacecraft was busy with other things, it took a total of 18 days to send all the images and other data back to the DSN.

Although it only spent a matter of minutes in the vicinity of Mars, and had very little in the way of instrumentation, Mariner 4 still managed to demolish the myth of Mars as an inhabitable, Earth-like world. It established the planet's atmosphere to be much thinner, and its surface temperature much lower, than scientists had believed up to that point. In fact, Mariner 4 made Mars look even more depressingly barren than it actually is. By chance, those few close-up pictures happened to focus on a rather untypical part of Mars, which was heavily cratered like the surface of the Moon.

As its numbering suggests, Mariner 4 was just one part of a larger programme of interplanetary missions. As well as several trips to Venus, the 1960s saw two other Mars flybys, by Mariners 6 and 7 in 1969, but these told scientists little more than Mariner 4 had. What was needed was a longer look, and that called for an orbital mission. This task fell on Mariner 9, which was duly launched in 1971. That was a particularly good year for Mars exploration, because the Hohmann launch window coincided with a Martian perihelion – something that only happens once every 16 years or so. When it does, the Hohmann trajectory is at its shortest, meaning the spacecraft gets to its destination quicker. Better still, it also means the spacecraft can carry a larger payload for the same amount of fuel. At almost a tonne, and loaded with scientific instruments, Mariner 9 was much larger and more sophisticated than its predecessors.

Unfortunately, Mariner 9 arrived at Mars at the worst possible time. The whole planet was engulfed in a gigantic dust storm, which completely obscured the surface. The robot probe was forced to wait in orbit from November 1971, when it arrived, until January of the following year before it could start taking photographs. But they were worth the wait – the first truly spectacular images ever seen of a world beyond the Earth. In all, Mariner 9 took more than 7,000 pictures over a period of nine months, totalling 54 megabits of data. Covering some 85 per cent of the planet's surface, at resolutions down to 100 metres, they revealed breathtaking topographic features like Olympus Mons and Valles Marineris.

During that same 1971 launch window, Russia sent two large probes of its own, Mars 2 and 3, each consisting of both an orbiter and a lander. The idea was for the orbiters to locate suitable landing sites for their respective landers. Unfortunately, that planet-wide dust storm put paid to that idea, and the landers had to go in blind. The first of them crashed, while the second appeared to land successfully, but then its radio went dead a few moments later.

The first successful landings on Mars had to wait until NASA's Viking 1 and 2 missions of 1976. These were sophisticated spacecraft, similar in concept to their Russian predecessors. They each consisted of an orbiter, to image the Martian surface and find suitable landing sites, and a lander. The latter used a heat shield and an aeroshell – an aerodynamically shaped protective casing – to enter the atmosphere, followed by a parachute which opened around 4.4 km altitude. In the thin Martian atmosphere, the parachute was only able to slow the decent from some 300 metres per second to about 60 – still much too fast to be survivable. So at 1.2 km altitude, the lander's terminal descent engine – similar to that of the Apollo lunar module – fired up to slow its fall to just 2 m/s on touchdown.

Both the Vikings landed successfully, and went on to send back some 10,000 images from the Martian surface. There were two possible data routes – either direct from the lander to the DSN on Earth, or relayed via the Viking orbiters. The latter route offered a higher data rate, but it could only be exploited when the orbiter was correctly positioned in the sky above the lander.

Despite the success of the Viking missions, they had one big drawback – they cost a lot of money. This may explain why NASA didn't go back to Mars for another 20 years … and when it did, it was with a deliberately low-cost mission. In 1976, Viking cost $2 billion; in 1997, the budget for the Mars Pathfinder mission was about a tenth of that (or even less, allowing for inflation).

While the Viking spacecraft had been over-engineered in every respect, Pathfinder was an exercise in KIS – 'Keep It Simple'. It was smaller than Viking – 270 kg compared to 570 – and it was just a lander, without an orbiter. Although most plans up to this point had included an orbiter, if the actual aim is to land on the surface, at the expense of some very complicated calculations, it's simpler and cheaper to switch straight from the Hohmann transfer orbit to a landing trajectory.

Another of Pathfinder's simplifications was to dispense with Viking's sophisticated landing engine. The first part of the descent used a parachute, just like Viking, to slow the lander to about 60 m/s. This was the point at which Pathfinder's cut-price approach began to diverge from that of its predecessor. At just a hundred metres or so from the ground, three simple rocket-assisted deceleration (RAD) motors fired for a few brief seconds – just enough to slow the impact to 14 m/s. That's a lot more than Viking's 2 m/s – and equivalent to a pretty nasty car crash, with g-forces in excess of 18 g. Pathfinder's trick was to envelop itself in airbags at the very last moment, resulting in a novel – if not entirely dignified – way to arrive on a new planet. The spacecraft bounced high in the air at least a dozen times, before finally coming to rest about a kilometre from its original touchdown site.

Despite its deliberately rough landing, the Pathfinder mission was a complete success. As a venerable old relic, it plays an important (if fictional) role in both the book and film versions of *The Martian*. It's the spacecraft that Mark Watney refurbishes to provide himself with a means of communicating with Earth.

Also making a cameo appearance in *The Martian* is Pathfinder's miniature rover, called Sojourner. This unusual name comes from a historical figure, Sojourner Truth, who was a civil rights campaigner in 19th-century America. It was suggested as a suitable name for a Mars rover by a young student following a competition – a naming method that has now become traditional for NASA rovers. Although Sojourner was tiny – just 10 kg in mass and 63 cm in length – it had its own instruments, camera and solar panel, and could communicate with the main lander at ranges up to 2 km.

Having a rover on the surface of Mars raises an important new problem – the time delay inherent in communications between Mars and Earth. This is less of an issue for static landers, where a delay between a command and its execution may be an irritation but can usually be worked around. But controlling a moving vehicle is another matter – especially given Mars's notoriously rocky terrain.

Radio signals travel at the speed of light, or approximately 300,000 km/s. As Earth and Mars move around the Sun in their respective orbits, the resulting time delay can vary from about three minutes at perihelion opposition to 24 minutes at aphelion conjunction. When Pathfinder landed in 1997, the Earth–Mars distance was about 190 million km, so signals took 10 minutes and 35 seconds in each direction. This meant that driving Sojourner by remote control was out of the question – just as it would be with any robot vehicle on Mars. It either has to be given tasks in discrete chunks, and then be left to get on with them, or else be completely autonomous.

NASA's next rovers were Spirit and Opportunity – or, more formally, Mars Exploration Rovers A and B – which arrived on the Red Planet in January 2004. At 180 kg, and 1.6 metres in length, they were much larger than Sojourner – and completely self-contained, without needing to rely on a separate static lander. Powered by their own large solar arrays, both rovers had a planned mission duration of 90 Martian sols, but they managed to exceed that by an enormous margin. Spirit lasted more than six years, covering 7.7 km and returning over 120,000 images before its systems finally failed. Opportunity fared even better, still going strong – and with over 40 km on the clock – at the start of 2017.

Spirit and Opportunity arrived on Mars using the same parachute/RAD/airbag system as Pathfinder. However, while this is a cheap and reliable method, it can't be used with larger or more delicate payloads. It might be suitable for landing stores and light equipment destined for a human mission, but certainly not the crew themselves.

A gentler and more sophisticated technique was used with NASA's fourth rover, Curiosity, when it arrived at Mars on 6 August 2012. Approaching a tonne in mass, and almost 3 metres long, Curiosity is comparable in size to a small city car. It employed the most sophisticated EDL system to date, returning to a fully powered descent like that of Viking – but with a couple of added twists, as shown in the accompanying illustration.

As usual, Curiosity entered the Martian atmosphere encased in an aeroshell and heat shield – but on a larger scale than anything that had previously been sent to Mars. In fact, it was slightly bigger than the Apollo command module, designed to return three astronauts from the Moon. Unlike earlier Mars missions, Curiosity performed a series of aerodynamic bank manoeuvres to achieve a controlled descent, like the Space Shuttle, before its parachute opened at 10 km altitude.

In another difference from previous missions, Curiosity used a downward-looking radar, called a terminal descent sensor, to locate a safe landing place free from large rocks and other potential hazards. Around 2 km altitude, when the rate of descent had slowed to about 100 m/s, the backshell and parachute detached themselves and the lander's powered descent vehicle (PDV) fired up. This slowed the spacecraft to a near-hover at 20 metres altitude – at which point the last, and most bizarre, phase of the landing took place. Just like a Skycrane helicopter on Earth, the PDV unreeled a tether to lower the rover gently to the surface. As soon as the PDV detected that the rover had touched down, it cut the tether, flew itself a safe distance away, and – its job done – deliberately crash-landed.

Entry, descent and landing of the Curiosity rover.
(NASA image)

Curiosity's official designation is the 'Mars Science Laboratory', which emphasises its primary function of doing serious science. The mission's main objective is to assess the ability of the Martian environment to support life – either now or, more likely, in the past. To this end the rover carries a whole arsenal of high-resolution cameras, geological tools and chemical instruments to help with the collection and analysis of rock samples.

Curiosity is powered by an RTG – a radioisotope thermal generator – rather than the solar panels used by earlier rovers. As explained in the previous chapter, an RTG isn't a nuclear reactor, but it does rely on naturally occurring nuclear reactions. This means it can produce much more power than a chemical battery of the same size. Curiosity's RTG contains about 5 kg of plutonium, which produces

2 kW of heat – some of which can be siphoned off and converted to electricity.

As with Spirit and Opportunity, Curiosity has two options when it comes to communications. Either it can talk to Earth directly, or messages can be relayed via one of NASA's orbiters, such as Mars Odyssey or the Mars Reconnaissance Orbiter. Both methods involve significant time delays – not simply because of the speed-of-light transit between Mars and Earth, but also due to bandwidth limitations and the need to wait for a clear line of sight between transmitter and receiver. For this reason, it's completely impossible for mission controllers to 'drive' Curiosity remotely from Earth. Instead, the rover's activities for each sol – or Martian day – are carefully planned in advance, and sent as a single batch of commands at the start of the sol. The rover is then left to carry out the prescribed movements, experiments and other activities by itself, before sending its data back at the end of the sol.

Lost in space

Half a century after Mariner 4, progress in Mars exploration can seem painfully slow. There are many reasons for this – the high cost of getting to the Red Planet, ever-changing political priorities, brief and infrequent launch windows, journey times measured in months rather than days. And to top it all, there's the Mars Curse. Spacecraft destined for the Red Planet have a depressing habit of getting lost in space.

Of the 43 Mars missions listed on NASA's website, no fewer than 17 were launched by the former Soviet Union (now Russia) between 1960 and 1988 – and not one of them could be called a success. Four failed to reach Earth orbit, three failed to leave Earth orbit for Mars, another four were lost en route to Mars. The other six were orbiters or landers which managed to get to the Red Planet, but then failed to carry out their designated mission for one reason or another. The closest the Soviets came to a success was Mars 3 in 1971 – that was the one where the lander touched down and transmitted for a few seconds before breaking down. The associated orbiter fared better – it continued to send back a trickle of data for eight months, though nothing to compare with the spectacular images Mariner 9 was taking at the same time.

Since the end of the Soviet era, Russia has only attempted two Mars missions – the Mars 96 orbiter-lander combination in 1996, followed by Phobos-Grunt in 2011. The latter had the ambitious aim of collecting a rock sample from the Martian moon Phobos (*grunt* is Russian for 'ground') and then returning it to Earth. But neither of these spacecraft made it out of Earth orbit, due to problems with their upper-stage rocket engines. The reasons for these failures – whether they were caused by hardware faults or software errors – is not entirely clear. In the case of Phobos-Grunt, the official failure report put the blame on shoddy engineering discipline: 'cheap parts, design shortcomings, and lack of pre-flight testing'.

NASA has had its failures too, although its overall record is much better. Only five out of 20 American launches destined for the Red Planet have ended in failure. Two of these were early Mariner missions, while the other three – Mars Observer, Mars Climate Orbiter and Mars Polar Lander – occurred in the 1990s. Only one of these five missions, Mariner 8 in 1971, failed to reach Earth orbit because of a problem with the launch rocket. The other problems all occurred en route to Mars, or after arrival, and seem to have been the result of design flaws in the spacecraft itself. The most notorious foul-up – not just among Mars probes, but in the whole history of space flight – must be Mars Climate Orbiter in 1999. The subsequent board of investigation found that 'the root cause for the loss of the spacecraft was the failure to use metric units in the coding of a ground software file used in trajectory models'. In other words, the software engineers measured rocket thrust using pounds instead of newtons.

For British readers, the best-known failure in the history of Mars exploration will be Beagle 2 – the shoestring-budget lander designed by Open University lecturer Colin Pillinger. This was a groundbreaking project – not so much for its scientific aims or engineering design, but for its business model. Up to that point, space exploration had relied on governments throwing large sums of money at it. Because the British government had little interest in doing this, Pillinger had to seek out a substantial amount of private investment too. In the end, more than half the funding came from sources outside government – proof that space exploration and private enterprise can go together. Not that Beagle 2 was a hugely expensive piece of kit: at an estimated £45 million, it was only a few per cent of the cost of NASA's Curiosity rover.

Beagle 2 hitched a ride to Mars aboard a European Space Agency (ESA) mission called Mars Express. Launched in June 2003, its main purpose was to deliver an orbiter loaded with scientific instruments – which successfully entered Mars orbit on Christmas Day of the same year. At the same time, the 33 kg Beagle lander detached itself from the main spacecraft and headed down to the surface of Mars … and was never heard from again. Its fate remained a mystery for 12 years, until images taken by NASA's Mars Reconnaissance Orbiter showed Beagle 2 sitting on the surface of the Red Planet. It's still in one piece – which came as a surprise to many people – but one of its solar panels seems to have blocked the radio antenna, preventing communication with Earth.

It's not clear whether Beagle 2's cost-cutting philosophy was a major factor in its failure, but it seems possible. The large sums of money that are normally pumped into space missions don't just buy the hardware – they also pay the salaries of designers, mission planners and software engineers, as well as all the necessary ground support, testing and evaluation. Attempting to minimise costs may cut corners on activities that looked like dispensable luxuries, but turned out not to be. That's what the Russians said happened with Phobos-Grunt, and it may explain Beagle 2's failure as well.

The loss of Beagle 2 was a serious dent to Britain's national pride, and to Europe's too. It would be another 13 years before ESA made a second attempt to land on the Red Planet – this time with a completely different type of vehicle. Called Schiaparelli, after the 19th-century 'discoverer' of Martian canals, it was primarily designed to test a sophisticated EDL sequence similar to that used by NASA on its Viking landers of the 1970s. Schiaparelli's scientific payload was minimal, and its sole power source was a battery that would only last a few days. That was okay, however, because its main function was to prove ESA knew how to land a spacecraft on Mars.

Schiaparelli began its descent on 19 October 2016. As with Viking 40 years earlier, it used first a parachute and then a retro-rocket to bring it to a carefully controlled landing on the Red Planet. Unfortunately, Schiaparelli's software seems to have decided it was safely down on the surface when it was still a couple of kilometres up in the Martian atmosphere. The spacecraft appears to have switched off its engine and fallen the rest of the way – which is one way to land on Mars, but not the one ESA was trying to demonstrate.

What lessons do all these tales of failure have for future human missions to Mars? If the statistics show anything, it's that missions have the best chance of success when they're carried out by an experienced team that is building on something they've done before. NASA hasn't had a single failure since 1999, and missions like Spirit and Opportunity have exceeded expectations enormously. That's in stark contrast to the Russian and European failures in the same period.

The statistics make depressing reading for any would-be newcomers to Mars exploration – but there's one striking counter-example. The Indian Space Research Organisation achieved a resounding success with its first interplanetary mission – the Mangalyaan Mars orbiter. It arrived in Mars orbit in September 2014, and was still fully functional two years later – a feat that not even NASA's engineers could claim with their first orbiter, Mariner 9.

Anything a human can do?

At least one robot probe has been dispatched towards Mars at virtually every launch window for the last 20 years (the single exception being 2009), and this trend is likely to continue for the foreseeable future. After a small static lander in 2018, NASA is planning to send another Curiosity-sized rover to the Red Planet in 2020 – using much the same technology as Curiosity, but with different science goals. 2020 will also see ESA and the Russian space agency Roscosmos teaming up to land the ExoMars rover, similar in size and complexity to NASA's Spirit and Opportunity.

Based on current plans, 2020 will see a lot of spacecraft heading for Mars – potentially more than any previous launch window. Besides NASA and ESA, the Indian space agency is also planning a follow-up to their Mangalyaan orbiter, possibly including a small lander and a rover this time. Other countries working towards a 2020 Mars mission include China and Japan, with both countries also thinking in terms of a lander/rover as well as an orbiter. China has already put a sizeable lander on the Moon – the 1.3-tonne Chang'e 3, which touched down in December 2013 and went on to deploy an Opportunity-sized rover called Yutu. The same team behind Chang'e 3 and Yutu are aiming to repeat that accomplishment on Mars.

The Japanese offering, called MELOS, is equally ambitious. Its rover would be smaller than Spirit and Opportunity, but precision-deployed by the same 'sky crane' technique as the much larger Curiosity. The preferred landing site is inside the Valles Marineris – Mars's Grand Canyon. As an optional extra, the MELOS proposal also includes the very first Martian aircraft, in the form of an unpowered glider. With a wingspan of 1.2 metres and a mass of 2.1 kg, it would be released during the lander's descent at an altitude of 5 km. Its estimated flight duration is four minutes, during which time it would travel 25 km over the surface of Mars, taking photographs as it goes.

The idea of flying an aircraft in the thin Martian atmosphere has been around for some time. As a problem in aerodynamic design, it's challenging but not impossible – similar to flying on Earth at 35 km altitude. An aircraft has the advantage that it can cover a large distance in a short time, and capture detailed photographs that couldn't be obtained in any other way. NASA has been considering the possibility of a glider-only mission to Mars – without the need for a separate soft-lander – since the 1990s. The glider would be deployed from an aeroshell as it descended under parachute, and with careful design might be able to stay airborne for up to an hour. The aircraft could even be solar-powered to increase its duration still further. Although not currently in NASA's Mars schedule, such missions could take place sometime in the 2020s.

Conceptual design for a Mars aircraft.

(NASA image)

Static landers, rovers, maybe even aircraft – if robots can explore Mars on their own, is there really a need for humans to go there? This is one of the great debates of space science, with pros and cons on both sides. Robots don't have to worry about food and life support for the long journey, and they're not bothered if it's a one-way trip. That reduces the total mass that needs to be sent to Mars, and when it comes to space travel lower mass means lower cost. And robots have other advantages besides financial ones. A human mission might only spend a month on Mars – it certainly couldn't compete with Opportunity's record of over 13 years' active work on the Red Planet. When it comes to purely scientific objectives, it could be argued that robots have the edge over humans. Yet that's not always true. Robots have their limitations.

On Earth, robotic vehicles fall into two broad categories: autonomous and remotely operated. The latter have a pilot or driver just like a conventional vehicle, except they're sitting somewhere else – not in the vehicle itself. The flying drones that are increasingly being used for aerial photography are a well-known example. This is never going to work on Mars, because of the time delay – so the only alternative is an autonomous robot. 'Autonomous' in this context doesn't mean the robot thinks for itself, just that it has a self-contained computer program, loaded with thousands of 'if-then' instructions, which tells it exactly what to do. This sort of technology is constantly improving, because there's a huge demand for it. Not for Mars rovers, of course, but for dozens of civil and military applications ranging from drones to driverless cars. Future Mars rovers will be able to benefit from that technology as it's developed.

Nevertheless, there will always be things humans can

do that robots can't. For example, the Curiosity rover has a whole range of sophisticated instruments for chemical analysis, but it's still no substitute for white-coated technicians in a laboratory on Earth. A solution to this problem, still using nothing but robots, would be a sample-return mission. Russia carried out three such missions from the surface of the Moon in the 1970s, but the most that was ever returned to Earth was 170 grams – a paltry amount compared to the 100+ kilograms collected by the last Apollo mission in 1972.

No one has ever attempted to return a sample from Mars, although 2011's failed Phobos-Grunt mission was supposed to collect one from Mars's moon Phobos. The moon's small gravity would have made the job relatively simple. A similar feat was performed in 2005 by the Japanese spacecraft Hayabusa, which collected a microscopic sample from a small asteroid and returned it to Earth five years later. But that asteroid was so small it had virtually no gravity of its own, so landing and taking off again was easy. With a surface gravity over a third of Earth's own, Mars would be a very different matter.

Roscosmos has a plan for a follow-up to Phobos-Grunt called Mars-Grunt, which if approved might take off some time in the 2020s. Mars-Grunt would be a small-scale mission, involving a tiny lander of just 20 kg. This would collect a few hundred grams of soil at the landing site, then return to the main spacecraft, waiting in orbit for the flight back to Earth. A more sophisticated plan is being considered by NASA, although this too is lacking formal approval. The NASA plan would involve multiple missions over a period of several years. To start with, a rover would be sent to collect and store rock samples – a role that might be fulfilled by the Curiosity follow-on planned for 2020. A second mission two years later would land near the first rover with an ascent vehicle and a small rover of its own, which would pick up the samples for return to Mars orbit. But that still isn't the end of it – yet another spacecraft would be needed to bring the samples back from Mars orbit to Earth.

This is beginning to sound as complicated as a human mission. In fact, the transition from robots to humans probably won't be made in a sudden jump, but as a gradual step-by-step evolution. A sample return mission is one possible intermediate option. Another is to use robots on the surface of Mars, but have humans controlling them from Mars orbit. This simplifies the human mission, by avoiding the need for a surface habitat and a large ascent vehicle. It also avoids the time delay problem, allowing robots to be operated in real-time, just as an operator sitting on an oil platform or a ship can control a robot submarine.

This approach has potentially huge advantages over the current system, in which a rover is loaded with a pre-packaged set of instructions at the start of each day. As former NASA astronaut Buzz Aldrin told the BBC in 2013: 'One programme manager, who was in charge of doing that with two robots for five years, has said we could have accomplished just as much in a single week, if we had had human intelligence controlling them from nearby – from an orbit around Mars itself'.

The Mars Piloted Orbital Station, or MARPOST, is a Russian proposal for a hybrid human–robot mission of just this type. Following an opposition-class, short-stay mission profile, the human crew would spend about 30 days in Mars orbit, during which time they would control a number of robots on the surface. Some of the samples collected would be flown back up to orbit, to be returned to Earth with the crew. Unfortunately, although the MARPOST idea has been around since the 1990s, it has yet to become a funded programme. A similar proposal was made by the American company Lockheed Martin in 2016. Called Mars Base Camp, it was accompanied by a timeline indicating that it would be feasible by 2028.

There is a distinct sense of déjà vu here. When you look back at proposals for the human exploration of Mars – whether from the 1950s, the 1970s, the 1990s or right now – they always seem to show something exciting happening around 10 to 12 years in the future. With the possible exception of the current ones, none of these optimistic predictions came close to fulfilment. This can lead to the cynical attitude that it's all too difficult, and it will never happen. Yet back in the 1960s something very similar *did* happen – when Project Apollo went from a standing start to the Moon in less than a decade.

FROM A SMALL STEP TO A GIANT LEAP

4

The race to the Moon

> Recognising the head start obtained by the Soviets with their large rocket engines, which gives them many months of lead-time, and recognising the likelihood that they will exploit this lead for some time to come in still more impressive successes, we nevertheless are required to make new efforts on our own. For while we cannot guarantee that we shall one day be first, we can guarantee that any failure to make this effort will make us last.

Those words, addressed to the United States Congress, were spoken by President John F. Kennedy on 25 May 1961. Six weeks earlier, America had lost the race to put a human into space when the Russian cosmonaut Yuri Gagarin orbited the Earth in Vostok 1. When its American equivalent Mercury 3 took astronaut Alan Shepard into space soon afterwards, it was only a much shorter suborbital flight. Kennedy knew America was on the back foot, and he was determined to win the next race. It would be a race with a clear goal, too, as he made clear in that same address to Congress:

> I believe that this nation should commit itself to achieving the goal, before this decade is out, of landing a man on the Moon and returning him safely to the Earth.

Kennedy didn't make this pronouncement on a whim. He'd taken advice from experts at NASA, who assured him the Moon was an achievable goal within the time frame. There were already plans for a three-seat successor to the Mercury capsule, and it even had a name – Apollo. It might have been used in various ways, but now all NASA's attention was focused on just one of them – the Moon landing. Kennedy's end-of-decade deadline was a tight one, and there was no time for messing around. By the middle of 1962, the basic mission architecture had been agreed.

In its final form, the Apollo spacecraft would consist of three modules, only one of which – the two-person lunar module – would land on the Moon. This consisted of two stages, the lower one with a descent engine and landing legs, the upper with a pressurised crew compartment and ascent engine. For most of the trip to the Moon and back, the three astronauts would live inside the more spacious command module. It was there that one crew member would remain in orbit around the Moon while the other two descended to the surface. Attached to the rear of the command module for the whole mission, barring the final re-entry into Earth's atmosphere, was an unpressurised service module, containing most of the spacecraft's support equipment as well as the large rocket engine needed to enter and leave lunar orbit.

The total mass of the Apollo spacecraft was some 45 tonnes – significantly larger than anything that had previously been put into Earth orbit, let alone sent to the Moon. To get it off the ground needed the world's largest launch rocket, by quite a margin. The Saturn V that sent Apollo to the Moon stood more than 100 metres high, and weighed 3,000 tonnes. Half a century later, it remains the largest and most powerful rocket ever built.

The Saturn V was essentially the brainchild of one man, Wernher von Braun, who was more convinced than any of his contemporaries that a journey to the Moon was an achievable goal. It was a vision he'd had at least as far back as the 1930s, when the Nazis were in power in his native Germany. Sadly, this reference to the Nazis isn't just an incidental detail; it's a key part of the Apollo story. It's an uncomfortable one, too, in a world that likes to keep its heroes and monsters in neatly separated compartments. But facts are facts – and Wernher von Braun was a Nazi.

Von Braun's first successful rocket was the V-2, which could reach the edge of space if fired directly upwards. It was hardly ever used this way, however. In the closing stages of the Second World War thousands of V-2s, each loaded with a tonne of high explosive, were fired against civilian targets in the cities of London and Antwerp. Falling to Earth at supersonic speeds, the attacks came without any warning, and innocent people were killed in their thousands. It would be comforting to think, just as von Braun himself maintained in later years, that he was simply an innocent scientist who saw his work twisted to destructive ends by the German military. Yet it's clear he remained wildly enthusiastic about the V-2, regardless of the uses it was put to. He became a member of the SS paramilitary organisation, rising to the rank of

Wernher von Braun dwarfed by the Saturn V's main engines.
(NASA image)

Sturmbannführer (equivalent to an army major), and he knew that slave labourers were employed in the V-2 factories.

Von Braun should have been put on trial as a war criminal. He wasn't, for the simple reason that his status as the world's number one rocket scientist made him too valuable.

He transferred his allegiance from the German army to the American one, eventually moving – along with most of his German colleagues – to America's 'Rocket City' in Huntsville, Alabama. It was there, in the early 1950s, that he designed the Redstone ballistic missile – a version of which would blast the Mercury 3 capsule, and Alan Shepard, into space in May 1961. By that time, von Braun had already started work on the Saturn family of launch vehicles   the first rockets to be designed right from the start as space launchers rather than as military weapons.

Getting to the Moon entails a lot more than coming up with a plan, building the necessary hardware and flying the mission. Perhaps in science fiction you could get away with that, but not in the real world. There are too many unknowns, too many things that have never been done before, too many ways it could all go pear-shaped. So NASA came up with a whole new space programme, called Project Gemini, simply to test out all the new technology and procedures needed for the Moon. The Gemini spacecraft was essentially a two-seat version of the Mercury capsule, and in ten flights between March 1965 and November 1966 it was used to test everything from spacesuit designs and fuel cells to docking techniques, orbital manoeuvres and the ability of astronauts to live in space for a week or longer.

The lightning-fast progress NASA achieved during the 1960s was far beyond anything the aerospace industry has seen before or since. Part of the reason, of course, was that the government was throwing money at the space programme as if there were no tomorrow. NASA's own estimate of the total amount it managed to spend between 1961 and 1973 is $20 billion. But there's another factor, too. The Americans weren't just *going* to the Moon – they were in a *race* to the Moon. Which begs the question, how were the other side doing?

At first sight, the Russians dominated the early part of the space race. They might not have had Wernher von Braun working for them, but they acquired a number of German engineers who had worked on the V-2, as well as V-2 hardware and production facilities. That was enough to get the Soviet space programme started – and it kicked off with a series of spectacular firsts. After Vostok 1 put the first human in orbit, Vostok 2 saw a person spend more than a day in space for the first time. Vostok 3 and 4 flew in space simultaneously – the first time that had been done. Vostok 5 and 6 was another dual mission – with Vostok 6 carrying the first woman into space. A modified Vostok called Voskhod 1 carried three people into space – the first multi-person crew. There were only two people on board Voskhod 2, but one of them stepped outside the capsule in the world's first extravehicular activity, or EVA. That was in March 1965, less than four years after Yuri Gagarin's original flight in Vostok 1.

The record looks good … but it's the tip of a lumbering iceberg, hiding a vast, semi-chaotic bureaucracy below the surface. Unlike the Americans, the Russians had no clear, cohesive strategy for getting to the Moon – or anywhere else in space. Impressive as they were, the Vostok and Voskhod missions were planned on a flight-by-flight basis – 'What *can* we do next?' rather than 'What do we *need* to do next?'

Viewed in its historical context, the space race was a facet of the larger 'Cold War' rivalry between Russia and America, which in turn had its roots in the battle between communism and free enterprise. There's an irony here, because the way the two countries approached the space race is almost diametrically opposed to their stated ideologies.

The NASA of the 1960s was a rigidly hierarchical, centrally controlled, government-funded organisation which did its job well if rather soullessly. Surprisingly, the Russians had nothing like it at the time. Plans for spacecraft and space missions came from two competing 'design bureaus', headed by the engineering equivalent of prima donnas who didn't see eye to eye at all. Functionally, the design bureaus behaved like private companies – which is exactly what they became, after the fall of the Soviet Union.

America's aerospace companies toed the line and built exactly what the government – NASA – told them to build. On the other side of the world it was different. The Soviet government adopted a hands-off attitude – they had more pressing priorities – and left the design bureaus to their free-for-all squabbling. One of the bureaus wanted to land on the Moon as soon as possible, just like NASA, while the other felt the near-term focus should be on a flight around the Moon, without an actual landing. For its own part, the government wasn't too comfortable with the Moon at all. Its main preoccupations revolved around economic productivity and military power, and if it had an interest in space at all, it was in how best it could be used to those ends. From this point of view, a large space station in Earth orbit, with the unique strategic perspective it could provide, seemed far more attractive than a trip to the Moon. It was against this backdrop of conflicting objectives that the Soyuz programme emerged in the mid-sixties. From the start, it was a spacecraft without a clear role – and one that ended up being pulled, by various stakeholders, in three or four directions at once.

In hindsight, it's difficult to imagine US intelligence wasn't aware of this turmoil in the Russian space programme.

But American paranoia tended to view anything to do with communism as a bigger and more immediate threat than it was. Maybe they thought the apparent confusion was a smokescreen designed to conceal just how close the Soviets were to a Moon landing. For whatever reason, the truth never made it through to the rocket scientists at NASA, who remained convinced they were in a close-run race with the Russians. This gave the Apollo programme an overwhelming sense of urgency, which ultimately led to its success ... but only after it had first led to its biggest disaster.

After several automated test flights, the first crewed Apollo mission was scheduled for launch in February 1967. It was a preliminary test of the CSM – the command and service modules – which would take three astronauts into Earth orbit. But it never made it off the ground. A month before the actual flight, with the spacecraft installed on the launch pad, the crew were shut inside for what should have been a routine ground test. The capsule was pumped full of pure oxygen – not at low pressure, as it would be in space, but at higher than atmospheric pressure to test for leaks. That was the first mistake.

Anyone who has been on a fire safety course will be aware that oxygen is one side of the notorious fire triangle. Another is 'fuel', meaning anything that will burn in the presence of oxygen. There was plenty of that on board – partly because no one had thought to build the spacecraft using fire-resistant materials, and also because this ground test involved a lot of extra paperwork inside the crew compartment. That was the second mistake.

The third side of the fire triangle is heat – a source of ignition to start the fire. An electrical spark will do here, and the Apollo capsule had enough poorly insulated electrical equipment to make such a spark not just a possibility, but – in hindsight – a virtual certainty. This was NASA's third, final, fatal mistake. At 6.31pm on 27 January 1967 the Apollo 1 crew reported a fire inside the command module. Less than a minute later all three astronauts were dead.

The fire was the biggest shock in NASA's history. From a 'make it happen' point of view they'd always done everything right. They'd designed a whole suite of innovative new technology to do what needed to be done as efficiently as possible. They'd set up a carefully planned programme of tests to make sure every component functioned the way it was supposed to. But that wasn't enough – they'd missed out an important step. They'd never done a proper risk assessment – they hadn't thought through all the things that *might go wrong*. Apollo 1 changed that. Overnight, NASA became one of the most risk-averse organisations in the world.

The immediate effect of the fire was to put the Moon race on hold – at least as far as the Americans were concerned. They had to redesign the Apollo spacecraft – and review the whole lunar mission – in a new, safety conscious way. This gave the Russians a chance to catch up, to some extent, although their Moon plans were still chronically disorganised. Their biggest triumph came in September 1968, when they sent a modified Soyuz capsule, called Zond 5, all the way round the Moon before bringing it safely back to Earth. There were no humans on board, but it did have a 'crew' of sorts – two tortoises, which survived the trip. Having beaten America's hare, the tortoises became the Soviet Union's unlikeliest heroes.

Zond 5 turned out to be Russia's last space 'first' of the sixties. They never managed to follow it up with a round-the-Moon Soyuz carrying a human crew. This was partly down to technical problems, but as much as anything it was a lack of political will. If they'd gone at it as whole-heartedly as the Americans, there's little doubt they would have succeeded. But it would only have been a flying visit, not a landing, which they knew the Americans would achieve before they could. The Russians were simply too far behind in the race, and to all intents and purposes they gave up trying to compete.

The mission that should have been Apollo 1 turned out – after another series of automated test flights – to be Apollo 7. The first flight of the CSM carrying a human crew, it blasted into Earth orbit in October 1968. It was followed on a ten-week drumbeat by three more shakedown flights – progressive tests of the CSM and lunar module, first in Earth orbit and then in lunar orbit – before the go-ahead was given for the landing mission. Apollo 11 was launched on 16 July 1969, and four days later Neil Armstrong and Buzz Aldrin became the first humans to set foot on the Moon. Four months after this, in November 1969, the feat was repeated by two more astronauts on Apollo 12.

Back in 1961, President Kennedy had given NASA the goal 'before this decade is out, of landing a man on the Moon and returning him safely to the Earth'. It was a goal they met, and not just with one man but four. NASA had won the race to Moon – but what would it do for an encore?

Mars is harder

For many people, Mars was the obvious next step after the Moon. At first sight it's an easy step, too. Mariner 4 swung past Mars in 1965, a mere six years after the Russians'

Luna 1 became the first spacecraft to reach the Moon. The two probes were comparable in size, and both were sent on their way by similar-sized rockets. Amazingly, if the super-economical Hohmann route is used, a Mars fly-by requires slightly less delta-v than a similar flight to the Moon.

Does that mean Mars is just as easy to get to as the Moon? That might be true if spacecraft trajectory was the only consideration, but there's another factor in the equation – mass. When there's a crew on board, a Mars-bound spacecraft needs a lot more mass than one that's only going to the Moon. People need life support, and that means taking along enough air, water and food for the entire mission. For an excursion to the planet's surface, a flimsy lander like the Apollo lunar module would never make it through the Martian atmosphere – it would need to be encased in a protective aeroshell and heat shield. To get back into space, against the pull of Martian gravity, also calls for a much larger rocket than the lunar module's ascent stage. All in all, this adds up to a lot more mass than Apollo had to take to the Moon.

On top of these basic functional issues, there's the comfort and sanity of the crew to think about. A mission lasting two years or more is going to need a much roomier vehicle than a week-long trip to the Moon. A convenient way to measure 'spacecraft spaciousness' is in terms of interior volume per person. The three-seat Apollo command module offered just two cubic metres per person. In contrast the Skylab space station, launched six months after the last Apollo mission to the Moon, offered its crew of three more than a hundred cubic metres per person. That's the sort of space astronauts would need on a Mars mission. It comes with a penalty, though – more space means more structure, and more structure means more mass. When linked up with the Apollo CSM, as shown in the following illustration, Skylab increased Apollo's mass by a factor of ten.

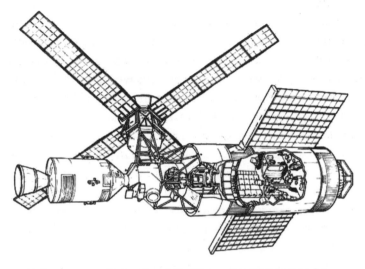

A spacecraft destined for Mars would need to be similar in size to the Skylab space station, seen here docked to the Apollo CSM (at the left of the picture).

(NASA image)

According to Tsiolkovsky's rocket equation, the amount of propellant needed to produce a given delta-v, assuming engine efficiency remains the same, rises in direct proportion to spacecraft mass. If the mass is ten times greater, then ten times as much propellant is needed to achieve the required delta-v. Sending a human expedition to Mars simply couldn't be done with a single Saturn V launch, as the Apollo Moon missions were. It's more likely to require three or four such launches – and even then, the third stage might need to be replaced with something more powerful, like the NERVA nuclear rocket. Indeed, the number of rocket launches would be even greater if the available technology was limited to present-day launch vehicles, rather than massive rockets in the Saturn V class.

Long before the crewed spacecraft leaves Earth, all the essential equipment would have been sent ahead on economical Hohmann trajectories – probably during the previous launch window two years earlier. This would include a surface habitat, one or more surface rovers, food and life support systems, and scientific equipment. Most important of all is the Mars ascent vehicle, or MAV, which is going to carry the crew back up to Mars orbit at the end of their stay. The MAV needs to be a substantial spacecraft in its own right.

Taking off from the surface of Mars doesn't require the same sort of multi-stage launch vehicle needed to get to Earth orbit, because Martian gravity is weaker than Earth's. The delta-v to get from the surface up to Mars orbit is about 4 km/s – less than half the Earth equivalent. Nevertheless, it's still twice the delta-v required for a suborbital space flight on Earth, such as that of Mercury 3, back in 1961. So getting a crew off the surface of Mars isn't a trivial feat.

Only when the MAV is safely in place would the astronauts themselves leave Earth. As already explained, the mission is going to need a big spacecraft – much too big to lift off from Earth in a single piece. Instead, the crew transit vehicle will have to be put together in Earth orbit, like a small space station – and one that's equipped with its own interplanetary propulsion system. One possibility here is NASA's proposed Deep Space Habitat (DSH), based on a small number of ISS-derived modules coupled to either a chemical rocket or an ion drive. When completed, the DSH might be comparable in size to Skylab – a cylinder 25 metres long and 6 metres in diameter.

Another thing the DSH will need, which wasn't too much of a problem for Apollo, is protection against high-energy radiation. The use of the word 'radiation' in this context can be slightly misleading. The term is most commonly used to refer to electromagnetic radiation, which consists of streams of photons moving in a straight line at the speed of light. That sort of radiation encompasses everything from the 'radiation hazard' of a microwave oven or an X-ray machine to the fearsome gamma rays unleashed by a nuclear explosion. It also includes more benign forms of radiation such as light itself. But the radiation that causes problems in interplanetary space is a different thing altogether.

Here the radiation in question consists of streams of charged particles – mainly protons and electrons. Some of these, called 'cosmic rays', originate outside the Solar System, but the majority come from the Sun itself – the solar wind. These particles are potentially very harmful, causing tissue degeneration, neurological effects and direct damage to DNA. The danger is particularly high during periods of increased solar activity, when large numbers of highly energetic protons are ejected from the Sun. Fortunately, because all these particles are electrically charged, their trajectories can be bent by a magnetic field (for readers old enough to remember, this was how pictures were formed on a television screen back in the days of cathode ray tubes). The Earth's own magnetic field deflects almost all the incoming radiation, keeping us safe down here on the surface of the planet. Because the field extends well out into space, it protects astronauts in low Earth orbit too, such as those aboard the

ISS. Much further than that, though, and radiation starts to be a serious worry.

To date, the only humans who have been fully exposed to the solar wind are the Apollo astronauts who went to the Moon. In that case, however, the exposure only lasted a few days, and the radiation dose they received stayed well within acceptable limits. That would not be true if the same level of exposure continued for several months, as it would on a trip to Mars. The physiological effects of over-exposure – 'radiation sickness' – are well-known, but recent studies have suggested that adverse psychological effects may kick in even sooner. Small levels of neural damage may impair an astronaut's ability to make rapid, complex decisions – an ability which would be critical if an emergency arose in deep space.

So any spacecraft venturing out into interplanetary space – NASA's DSH, for instance – will need an effective form of radiation shielding. One possibility is to set up an electromagnetic shield, like a miniature version of the Earth's own. But that would mean a continuous drain on the ship's power, and there are simpler alternatives. The phrase 'radiation shielding' may conjure up visions of thick plates of lead – but that comes from thinking of radiation in terms of X-rays or gamma rays. For the sort of charged-particle radiation we're talking about here, a much lighter form of shielding would suffice – for example a plastic material such as polyethylene, or even a layer of water.

Of course, it might not be possible to spare that much water, since it would be in demand for other things like drinking and washing. As such it would be carefully hoarded and recycled, just as it is on the ISS. British astronaut Tim Peake put it bluntly in an interview in October 2016: 'Yesterday's pee is this morning's coffee basically, but actually it tastes absolutely fine'. This raises another question, though – what about solid waste? It's been suggested that, as it builds up during the journey, human excrement would be an ideal substance to add to the spacecraft's radiation shielding (and when the astronauts get to Mars, it will make great fertiliser if they want to grow their own food – as Mark Watney discovered in *The Martian*).

So it shouldn't be too difficult to set up some form of radiation screen to provide day-to-day protection. An all-out solar storm would be a different matter, though. To deal with that contingency, the DSH would need to have some form of highly shielded 'storm shelter' built into it.

There's another big difference between a trip to Mars and a trip to the Moon – the degree of isolation. The Apollo astronauts could communicate with their Earth-bound colleagues without significant delay, and in any case they were never in space for more than a few days. Even on a long-duration mission in Earth orbit, astronauts on the ISS can phone their friends and family back home, and indulge in perfectly normal conversations with no greater time delay that an ordinary long-distance call. If something goes drastically wrong on the space station, they know they can always get back to Earth, using one of the permanently docked Soyuz capsules, within a few hours. It wouldn't be like that on the way to Mars – the crew would be on their own, both in the social and physical sense. At some points of the journey, a speed-of-light radio transmission could take as long as 20 minutes to travel between the spacecraft and ground control. That means 40 minutes could slip by between asking a question – maybe on an urgent, life-or-death matter – and receiving a reply.

The ability of a crew to cope with the isolation of a Mars

flight doesn't need a space flight to study it – just a suitably isolated crew. That can be done more cheaply, and more safely, in an Earth-based simulation. One such exercise was Mars-500, carried out in 2010–11 by the Russian Institute of Biomedical Problems, in collaboration with ESA and the Chinese government. Located near Moscow, the experiment was designed to simulate a 520-day opposition-class mission, including 30 days on the surface of Mars. The latter was represented by a large, hangar-like section of the facility which was opened up at the appropriate time. For the rest of the mission, the crew were confined to a hermetically sealed 'spacecraft', consisting of a number of interconnected modules. Its total internal volume was 300 cubic metres, comparable to NASA's Skylab of the 1970s. But Skylab was designed for three astronauts, while Mars-500 had a crew of six. For this experiment, they were all male – three Russians, two Europeans and one Chinese.

Mars-500 was deemed to be a success, with no serious psychological problems or personality conflicts developing. The crew members seem to have got on surprisingly well with each other … though perhaps it's only surprising because we're so used to watching reality shows like *Big Brother*, where the participants are deliberately chosen so they *won't* get on with each other. With Mars-500, as with any real space mission, the actual selection process was the exact opposite of this.

Mars-500 looked at the short-stay option of an opposition-class mission. With a conjunction-class mission – which needs less rocket fuel – the journey times would be similar, but the stay on Mars would be much longer. This introduces new problems of its own, which were addressed in another Mars simulation, called HI-SEAS. That may seem a strange name for the representation of a famously dry planet, but it stands for 'Hawaii Space Exploration Analogue and Simulation'. It was funded by NASA, but administered by the University of Hawaii, where the experiment was situated. Its focus was not on the space flight itself, as in Mars-500, but on a long-duration stay on the surface of Mars.

The main HI-SEAS experiment lasted for exactly a year, from August 2015 to August 2016. As with Mars-500 there were six crew members – four Americans and two Europeans – but this time with a 50-50 male-female balance. All the participants were scientists and engineers with a professional interest in Mars – exactly the sort of people who might be chosen for a Mars mission. This is an important point. It's not enough to select a crew who are going to be psychologically stable and get on in each other's company – they should be able to do their job when they get to Mars, too.

The HI-SEAS crew spent most of their time inside their small, enclosed surface habitat. This is one of the dull realities of a long-stay mission, which is often glossed over by Mars enthusiasts and science fiction writers. The amount of time astronauts could spend outside, exploring the surface of Mars, would be severely limited by the need to wear heavy pressure suits with all their life-support trimmings. Even though they were on Earth, the HI-SEAS crew had to do the same whenever they needed to work outside. The location of the experiment, on the side of a volcano on the big island of Hawaii, was specifically chosen because the terrain is quite Mars-like. The sense of being on the Red Planet was further enhanced by restricting the participants' communications to email, and even that with a simulated 20-minute delay. Their food supplies were limited too, to reflect what might be available on a real Mars mission.

As with Mars-500, the main purpose of HI-SEAS was to test whether the crew could hold up psychologically under the simulated conditions for the required length of time. The experiment seems to have ticked that box successfully, although several crew members were bothered by the lack of privacy. However, this sort of simulation is likely to represent a worst-case scenario from the psychological point of view, because deep down the crew know they're only camping on the side of a Hawaiian mountain. On a real mission, the excitement and novelty of being on another planet would override many of the day-to-day annoyances of an Earth-bound simulation.

Risk management

The increased complexity of a crewed Mars mission – considering life support, radiation shielding, spacecraft size and the need for multiple launches – explains why the step from the Moon to Mars is so much bigger for humans than for robot probes. It's not just that more equipment must be built and more work must be done on mission planning, but there's also the matter of checking that every system and subsystem is going to function the way it's supposed to. In the context of the risk-averse and safety-conscious NASA that emerged in the wake of the Apollo 1 fire, this means going through a long, step-by-step process of technology validation and flight testing before a trip to Mars can even be considered.

The Apollo mission profile was much simpler than one to Mars would be, but it still went through a methodical process of incremental testing. The flight that everyone remembers – Apollo 11 – only happened after a long series of shakedown tests of each subsystem and procedure. The astronauts them-selves, many of them coming from a test-pilot background, understood the need for this as well as anyone. Popular media may portray them as risk-taking daredevils, but the opposite is closer to truth. A case in point is Apollo 10 – a mission that is forgotten by most people, and baffling to those who do remember it. By the time Apollo 10 was ready for launch in May 1969, Apollo 8 had already tested the CSM in lunar orbit, and Apollo 9 had tested the lunar module in Earth orbit. The programme called for one more flight before the actual landing mission – a test of the lunar module in orbit around the Moon.

But was that necessary? Based on media stereotypes, one might imagine the Apollo 10 crew pushing to be allowed to land, while over-cautious NASA management insisted on ticking that one last box on the checklist. In fact it was the other way around. Senior NASA administrators, still wor-ried that Russia might get to the Moon first, were desperate to cut to the chase and go straight for a landing. But the commander of Apollo 10's crew, Tom Stafford, knew they couldn't do that. It was too much of a risk – the spacecraft still had too many systems that hadn't been properly tested. He assured his bosses that if Apollo 10 was switched from a routine shakedown flight to a history-making Moon landing, 'this flight crew won't be on it'.

The purpose of testing, of course, is to minimise the chances that something will go wrong. But it's impossible to reduce this risk to zero. At any point during a space mission something *may* go wrong, and it makes sense to have plans ready in advance to deal with every imaginable contingency. In the jargon of rocket science, they're called abort modes.

The best-known case of an abort mode being invoked

happened in April 1970, three days into what was supposed to be the third Moon landing flight – Apollo 13. There was a huge explosion in the service module, which damaged both power and life-support equipment. Fortunately, the astronauts survived the explosion, and so did the radio connecting them with mission control on Earth. There was no way they could proceed with the lunar landing, but the mission plan had a convenient abort option built into it. This was a so-called 'free return trajectory' – carrying on to the Moon, then using the lunar gravity field to fling the spacecraft back towards the Earth. It meant the astronauts had to remain in space for another three days after the accident – which wasn't pleasant, due to the damaged heating and life-support systems, but at least it was survivable.

A similar accident on the way to Mars would be a different matter. The flight plan to Mars doesn't have the same range of convenient abort options as a lunar trip. It does offer the option of a free return trajectory, but this involves an incredibly slow journey home – much longer than a Hohmann trajectory. If a Mars mission suffered an Apollo 13-style accident at a similar juncture, it might be three years – not three days – before the crew were safely back on Earth. If the life-support systems were as badly damaged as they were on Apollo 13, it's doubtful the crew's physical and mental health would hold up throughout the long, depressing journey home.

Considerations like these make it even more important to ensure the Mars spacecraft is robust and resilient enough to survive a wide range of eventualities. There's no substitute for careful, step-by-step testing – of the spacecraft hardware, of the various phases of the mission, and of the crew themselves. That will take a lot of time, and a lot of money.

Does this explain why NASA didn't carry on to Mars after the resounding success of Apollo? In a word – no. Although it's only fiction, Stephen Baxter's 1996 novel *Voyage* shows very convincingly how NASA could have landed three people on Mars in 1986, if it had made a commitment to do so immediately after the Apollo 11 landing. The book isn't science fiction – it's an alternative version of history, based on technology that existed at the time the story was set. The spacecraft that goes to Mars is made from bits of Saturn V, bits of Apollo, bits of Skylab … and a few bits that in the real world went on to the Space Shuttle. But in Baxter's world, the Shuttle was cancelled, and all the money the government would have put into it – and into robot explorers like Pioneer and Voyager – pays for the trip to Mars. Unlike most fiction writers, Baxter doesn't gloss over the vast amount of technical development, flight testing and crew training such an undertaking would involve. He even includes a couple of Apollo 1-style setbacks, and still manages to make the deadline.

But, as we know, the reality was altogether different. Let's look at what actually happened.

BIG PLANS

5

Visionaries versus politicians

Even now science can detail the technical requirements for a Mars expedition down to the last ton of fuel. Our knowledge of the laws governing the solar system – so accurate that astronomers can predict an eclipse of the sun to within a fraction of a second – enables scientists to determine exactly the speed a spaceship must have to reach Mars, the course that will intercept the planet's orbit at exactly the right moment, the methods to be used for the landing, take-off and other manoeuvring. We know, from these calculations, that we already have chemical rocket fuels adequate for the trip.

These words – which are still perfectly sound from a technical point of view – were written by Wernher von Braun in 1954. That's the same Wernher von Braun who worked for the Nazis during the Second World War, and went on to design the Saturn V rocket in the 1960s. His 'Mars Project', initially conceived in 1948, was the archetypal scientist's pipe dream: massively ambitious, brilliantly conceived … and economically impossible. It called for dozens of rocket launches which would gradually put together a fleet of vehicles in Earth orbit – some of them carrying human crew, others purely for cargo. The fleet would follow standard Hohmann orbits to and from Mars, necessitating a stay of around 400 days on the Red Planet. The only impractical aspect of von Braun's plan was his EDL strategy, employing winged aircraft that would fly down through the Martian atmosphere and land horizontally like a Space Shuttle. That would never work – but only because Mars has a much thinner atmosphere than scientists believed at that time.

Politicians would never go for a budget-busting plan like von Braun's. But what about something with a nice, practical military dimension to it? Around the same time the Mars Project was conceived, some of the people behind America's hydrogen bomb proposed a novel alternative to the rocket. It was called 'nuclear pulse propulsion', and it derived its energy by exploding a series of nuclear bombs, one after the other. If properly designed, such a system could produce a similar thrust to a large chemical rocket, but with much greater efficiency. The ratio of payload to total mass, which is typically only 3 or 4 per cent for a rocket, might have been as high as 45 per cent with such a design. For propellant, it would have used a dense metal like tungsten, shaped in such a way that when it was vaporised by the explosion, most of the resulting high-energy gas went in the required direction.

Frightening as it may seem, the American government seriously considered nuclear pulse propulsion in the early days of the space race. To ensure it was exploited as quickly as possible, they set up Project Orion in 1958 with the motto

'Mars by 1965, Saturn by 1970'. The first goal – Mars – would require a huge nuclear-propelled vehicle to be constructed in Earth orbit using several conventional rocket launches. Once the spacecraft was ready, hundreds of small atom bombs would be exploded in quick succession to give it a very high delta-v. This would put it on a super-fast trajectory to Mars, allowing the entire round trip to be completed in just 125 days.

Project Orion came close to starring in one of the most memorable science fiction movies of the 1960s, *2001: A Space Odyssey*. Directed by Stanley Kubrick – and coming just four years after his nuclear-apocalyptic masterpiece, *Dr Strangelove, or How I Learned to Stop Worrying and Love the Bomb* – this was a joint project between Kubrick and science fiction author Arthur C. Clarke. As the latter recalled in his book *The Lost Worlds of 2001*:

> When we started work on *2001*, some of the Orion documents had just been declassified, and were passed on to us … but after a week or so Stanley decided that putt-putting away from Earth at the rate of twenty atom bombs per minute was just a little too comic. Moreover – recalling the finale of *Dr Strangelove* – it might seem to a good many people that he had started to live up to his own title and had really learned to Love the Bomb.

In retrospect, Project Orion raises the concept of an 'accident waiting to happen' to spectacular proportions. Fortunately, it was cancelled before it ever reached space – but not because the people behind it thought it wasn't going to work. In August 1963, the United States signed the first nuclear test ban treaty, the terms of which specifically prohibited the testing of nuclear weapons 'in the atmosphere, in outer space and under water'. Technically the atom bombs used in Project Orion weren't weapons, they were part of a propulsion system – but that kind of technicality wasn't going to convince anyone.

By the time Project Orion was cancelled, the focus of NASA's attention was already firmly on the Moon. Then in August 1969, just a month after Apollo 11, NASA's chief administrator suddenly announced a plan that – if supported by the American people – could see humans set foot on Mars just as Armstrong and Aldrin had done on the Moon. The Mars plan had strong echoes of Apollo, with a large composite spacecraft travelling from Earth to the Red Planet, followed by a landing by some crew members while at least one stayed behind in orbit.

Once again, the man behind this latest proposal was Wernher von Braun – his last major contribution before retiring in 1972 at the age of 60. It would be an opposition-class mission, employing a Hohmann trajectory on the way out followed by a short stay on Mars. The return flight would use a different trajectory, passing close to the planet Venus to get a gravitational boost on the way back. Making this work properly called for very precise timing. The crew would have to leave Earth orbit on 12 November 1981, allowing them to reach Mars on 9 August 1982. They would spend 80 days there, leaving on 28 October 1982 and finally arriving back on Earth on 14 August 1983.

The Mars spacecraft would be launched by a Saturn V, the same as Apollo, but with the Saturn's third stage replaced by NERVA – the Nuclear Engine for Rocket Vehicle Application that was described in Chapter 2. Despite the word 'nuclear', it's important to remember that NERVA wasn't some crazy

Dr Strangelove idea like Project Orion. It didn't involve any atom bombs – just a carefully controlled nuclear reactor to generate a constant supply of heat, the same way the reactor in a nuclear power plant or nuclear submarine does. NERVA was a perfectly feasible idea – as indeed was the rest of von Braun's plan, if the US government decided to pursue it. But would they?

The race to the Moon was all about politics with a big P – the clash of ideologies between capitalist America and communist Russia. Less abstractly, the early space race can be seen as a natural offshoot of the nuclear arms race that raged between the two countries. The first space launchers on both sides were repurposed intercontinental ballistic missiles. The message was obvious – if a rocket could put a space capsule into orbit, then the same rocket could send a nuclear warhead from one hemisphere to the other. But when it came to Mars, this 'big politics' didn't apply. By the time NASA was ready to start thinking about the Red Planet, the Russians had decided to focus solely on Earth-orbital operations. In October 1969, just three months after Apollo 11, the Soviet leader Leonid Brezhnev made his position perfectly clear:

> We are going our own way: we are moving consistently and purposefully. Soviet cosmonautics is solving problems of increasing complexity. Our way to the conquest of space is the way of solving vital, fundamental tasks – basic problems of science and technology. Our science has approached the creation of long-term orbital stations and laboratories as the decisive means to an extensive conquest of space. Soviet science regards the creation of orbital stations with changeable crews as the main road for man into space.

If the Soviets were out of the Mars race, America was less than wholehearted too. There was no Martian equivalent of the bullish 'man on the Moon' goal set by President Kennedy in 1961. His successor at the time of Apollo 11, Richard Nixon, was much more cautious about America's way forward in space. His first step was to commission a Space Task Group to evaluate the options. The team recommended that NASA's funding should stay at the Apollo level, and that its focus should be on the continuing development of human space flight with the ultimate aim of a von-Braun-style Mars landing in the 1980s.

This was all promising stuff – but Nixon rejected the suggestion out of hand. Instead, he cut NASA's budget and instructed it to deliver a more balanced programme of robotic and crewed missions. The latter would be focused around a completely new piece of technology, the Space Shuttle. A semi-reusable spaceplane, this was largely irrelevant to a Mars mission, and even to continued exploration of the Moon. Just like his Russian counterpart, Nixon had effectively confined his country's human space flight horizons to Earth orbit.

The exploration of Mars – and the rest of the Solar System – was left to robots. These were purely scientific missions; there was no pretence that they were the build-up to a crewed flight, in the way early lunar probes had paved the way to Apollo. Surprisingly, this emphasis on scientific goals introduced a completely new dimension to space politics. In place of the big politics of government, this was 'politics with a small p' – within the science community itself. Mars had to compete with the other planets, with the moons of Jupiter and Saturn, with asteroids and comets, even with the Sun. There were perfectly valid scientific reasons for studying all these objects, regardless of whether humans might ever want to visit them. Every object in the Solar System developed its own 'pressure group' – scientists whose careers centred around it, and who could offer countless arguments as to why it was the obvious next target for a space mission. Even with Mars itself, there was strong internal competition between teams with different priorities – orbiters, rovers, sample return missions – all lobbying for money alongside anything that might be aimed at eventual human exploration.

Politicians devote a huge amount of time to arguing about how taxpayers' money should be spent. What fraction of the total should be allocated to scientific research, and how much of the science budget should go on space? How should the space budget be apportioned between crewed and uncrewed missions? According to one school of thought, human space flight in any form is an utter waste of government funds. For many people – members of the general public, and even some scientists, as well as socially conscious politicians – taxpayers' money should always be spent here on Earth rather than in outer space. While there's some validity to this argument, it misses the point that *money spent on space travel doesn't disappear into outer space*. Most of it goes into creating jobs – and not just the jobs of astronauts and rocket scientists, but people in manufacturing, construction, software and services. In the ultimate analysis, a healthy space programme is good for the economy.

Nevertheless, space is a subject that often suffers from a poor public image – a fact that's particularly noticeable in comparison to the incredibly high profile it enjoyed during the 1960s. When President Nixon announced the Space Shuttle project, it's unlikely that he foresaw its most damaging effect – to transform the public's perception of human spaceflight from 'exciting' to 'boring'. In its 30-year career, it only made attention-grabbing headlines twice – on the two occasions that a Shuttle crashed, killing seven astronauts both times. That's an appallingly bad public relations record.

Mars Direct

The 1990s saw something of a renaissance in Mars exploration, with the first new missions since the Viking landers of the mid-seventies – but they were still just robot probes like Mars Global Surveyor, Pathfinder and Sojourner. As far as NASA was concerned, human expeditions to the Red Planet were so far in the future they fell off the right-hand-side of its project plans – even the most forward-looking ones.

One person who took exception to the situation was an aerospace engineer and vocal Mars advocate named Robert Zubrin. He was frustrated by the way NASA seemed to waste time and money over-engineering every project it took on. It had turned into an organisation of perfectionists – and everyone knows perfectionists never achieve anything. Part of the problem went back to the 'risk averse' culture NASA had adopted in the wake of the Apollo 1 fire. Another part came from trying to please a whole spectrum of special interest groups at once – both in the scientific and political communities – and not actually managing to please any of them. Zubrin was convinced there was a simple way to Mars, right there under NASA's nose, which it was steadfastly refusing to see.

Zubrin called his concept Mars Direct. Originating in the early nineties, the most detailed account of it can be found in his 1996 book *The Case for Mars*. At its simplest, Mars Direct just needs two rocket launches in consecutive Hohmann windows, slightly more than two years apart – though for reasons of safety and sustainability, a third launch in parallel with the second would be desirable. Each launch would send approximately 40 tonnes of payload to Mars, using rockets in the easily achievable Saturn V class. Instead of a NERVA third stage, the Saturns would get their extra delta-v courtesy of Shuttle-style solid rocket boosters.

The first launch dispatches an uncrewed spacecraft to the surface of Mars. Zubrin calls this the Earth Return Vehicle, or ERV – and as the name suggests, its eventual job will be to bring the astronauts back home. There's a good reason for sending it two years early – it's going to need most of that time to synthesise its own rocket propellant via in situ resource utilisation (ISRU). The ERV's engines are designed, rather unusually, to burn a mixture of methane and oxygen. Methane isn't a standard rocket fuel, but like kerosene it's a chemical compound of carbon and hydrogen … and a much simpler one to make, if you happen to be on the surface of Mars.

The Martian atmosphere contains plenty of readily accessible carbon and oxygen, in the form of carbon dioxide. The only ingredient that's missing is hydrogen, but that's a very light substance. Over a hundred tonnes of propellant can be made from just six tonnes of hydrogen – and the ERV can easily bring that, in conveniently compressed liquid form, all the way from Earth. The only other thing it needs is a power source, which in Zubrin's proposal takes the form of a small nuclear reactor. As soon as it gets to Mars, the ERV can power up its on-board chemistry set and start making all the methane and oxygen it needs.

Only when the ERV is fully fuelled up and ready for service will the second spacecraft – the one carrying the crew – be sent on its way in the second launch. To keep energy requirements to a minimum, Zubrin's plan calls for a conjunction-class mission, using Hohmann orbits on both outbound and return flights, with a long stay of 18 months on the surface of Mars. The idea, of course, is that the crew will land as close as they can to the predeployed ERV. Just in case things don't turn out that way, Zubrin proposes a third launch, a few days after the second, carrying a backup ERV. If everything goes smoothly, the backup won't be needed by this crew – so it can stay on Mars ready for the next set of astronauts, two years later.

The name 'Mars Direct' comes from the fact that the crewed spacecraft goes straight to the surface of Mars, without leaving anything behind in orbit. The crew use the ERV for the return journey – which again is a direct flight, all the way from Mars to Earth. There's a subtle difference here from the mission profile portrayed in *The Martian* – and in many other recent proposals – which use a single transit vehicle for both the outward and return flights, together with smaller descent and ascent vehicles between Mars orbit and the surface. But Zubrin also sets out a strategy for a variant of his plan along those more familiar lines – which he calls 'Mars Semi-Direct'. In this scenario it's the Mars Ascent Vehicle, or MAV, which is predeployed a couple of years early and fuelled up via ISRU.

Just like von Braun's Mars Project four decades earlier, Robert Zubrin's ideas were nothing more than words, numbers and diagrams on paper. But there were a couple of important differences. Zubrin's plan used already proven technology, and it looked to be affordable within a standard space-programme-sized budget. Its basic elements – the imaginative reuse of tried-and-tested technologies, ISRU, the split-mission strategy – suddenly started turning up in a whole range of Mars proposals, left, right and centre. Even NASA eventually decided to jump on to the bandwagon.

One step at a time

NASA's approach to space has always been to take one step at a time. There's nothing wrong with that – it's what got Apollo to the Moon. Each step of that project was part of a clear, coherent plan directed at that one single goal. Even the Gemini flights, which might look like an unnecessary diversion, had the sole aim of testing the various new procedures that would be needed for Apollo, from spacecraft docking to EVAs. When NASA was finally ready for the Apollo missions, it didn't jump straight to the Moon landing. The spacecraft had to go through a whole series of incremental tests first.

Although Apollo adopted a step-by-step approach, it never lost sight of the ultimate reason for its existence – putting humans on the surface of the Moon. It was a textbook example of *goal-driven* engineering. That's not the only option, though. It's entirely within the rules to design, build and test a new piece of technology before you've decided what you're going to do with it. If you're creating something that has never existed before, why put artificial constraints on its use? That's a philosophy that appeals to some people – the *innovation-driven* approach.

This approach involves taking one step at a time, too – but instead of steps towards a clearly defined goal, they're steps into the unknown. First develop the new technology, then see what can be done with it. That's what the Russians did with their Vostok and Soyuz spacecraft, and NASA followed in their footsteps when it built the Space Shuttle. Unfortunately, operating the Shuttle turned out to cost so much there was no money left to exploit it to its full potential. It wasn't meant to be an end in itself, but that's what it turned out to be: an expensive way of travelling without ever reaching a destination.

The Shuttle did belatedly find a niche for itself in the construction of the ISS. That was primarily a Russian initiative, though, not NASA's – a follow-on to the successful Mir space station of the 1990s. Nevertheless, the Russians welcomed the Americans' involvement for the money they could bring, and in the end the Shuttle played a crucial role in getting the ISS built on schedule. Just as the Shuttle ended up doing something useful, NASA's innovation-driven approach may one day get it to Mars – but it won't be via a directly focused engineering programme à la Apollo.

Mars enthusiasts understandably find the situation frustrating. The fact is, however, that NASA engineers are slowly but surely getting closer to putting humans on the Red Planet. Their increasingly sophisticated robot probes, such as the one-tonne Curiosity rover, are testing and de-risking a whole range of essential technologies and procedures. NASA's follow-up to Curiosity, an as-yet unnamed rover to be launched in 2020, will carry an experiment designed to test the feasibility of ISRU, which plays a key role in virtually all current plans for human missions to the Red Planet. Dubbed MOXIE – for Mars Oxygen ISRU Experiment – it's designed to extract pure oxygen from atmospheric carbon dioxide. If successful, the same process could be used on a larger scale to produce oxygen for astronauts to breathe, and for use as an oxidiser in the MAV's rocket engines.

Closer to Earth, even the ISS can be thought of as an important step in the same incremental process. Its very existence demonstrates the feasibility of undertaking a large-scale construction project in orbit, which is something that would need to be done to send humans to Mars. The ISS is made up of discrete modules, each launched separately and put together in orbit, just as a Mars spacecraft would be. That's not a trivial task in zero-gravity, carried out by a small number of construction workers clad in bulky space-suits and with only a limited range of tools at their disposal. Putting the ISS together ticked an important box on the Mars pre-flight checklist.

The International Space Station.
(NASA image)

The ISS has helped in another, more obvious way. Arguably the most daunting aspect of a human mission to Mars is its sheer length – at least 400 days for the round trip, if a short-stay opposition-class itinerary is followed. Earth-bound studies like the Mars-500 and HI-SEAS experiments described in the previous chapter are useful up to a point but without an Earth-orbiting space station like the ISS, it would be impossible to prepare astronauts for such a long-duration mission – or for physiologists to know whether it was a safe thing to undertake in the first place. A typical tour of duty on the ISS lasts six months – around 180 days – and many crew members have returned for a second, third or even fourth time. In 2016, American astronaut Scott Kelly and Russian cosmonaut Mikhail Korniyenko successfully completed an extended duration mission lasting 342 days. That wasn't the longest space flight, though – a total of four cosmonauts spent a year or more on board the ISS's predecessor, Mir. The record is held by Valeri Polyakov, with a continuous stay on Mir of 437 days. That's long enough for an opposition-class mission to Mars.

NASA itself makes an explicit link between the ISS and Mars. In October 2015, in a document called *Journey to Mars: Pioneering Next Steps in Space Exploration*, it made the following observation:

> The ISS is the only microgravity platform for the long-term testing of new life support and crew health systems, advanced habitat modules, and other technologies needed to decrease reliance on Earth.

'Microgravity' is the technical term for what is more often referred to as 'weightlessness' or 'zero-g'. That's the state commonly experienced by astronauts in space – not necessarily because they're free from any gravitational influence (the force of gravity is only a few per cent weaker at the altitude of the ISS than on the surface of the Earth), but because a spacecraft with its engine switched off is effectively in free fall. It doesn't matter if it's in orbit around the Earth, in orbit around the Moon or Mars, or coasting along on a Hohmann transfer trajectory – if gravity is all that's doing the driving, the astronauts inside behave to all intents and purposes as if they're weightless.

Microgravity represents an unnatural environment for human physiology, which evolved in the constant 1 g gravitational field of the Earth's surface. Nevertheless, experience on the ISS and its predecessor space stations has shown that, with appropriate exercise routines, the human body can adapt quite well to living inside a spacecraft. From a health point of view, the environment even has one big advantage over the Earth – after the first few weeks, there are no new cold or flu viruses to pick up. That's nothing to do with being in space *per se* – it's because the spacecraft's population is so small and self-contained (the crew of a submarine on a long tour of duty experience the same pleasant side-effect).

The NASA quote referenced above talks about 'technologies needed to decrease reliance on Earth'. These encompass state-of-the-art robotic devices and 3D printing technology, as well as advanced medical equipment and fire safety systems. Most important of all are the basic life-support functions that ensure the spacecraft's habitability over a long period. On the ISS these are provided by a complex Environmental Control and Life Support System (ECLSS), which maintains an Earth-like oxygen-nitrogen atmosphere, removes carbon dioxide and other poisonous gases, and recycles water as much as possible. Water is particularly critical, because it's not only used for drinking, washing and food preparation, but also – via a chemical process called electrolysis – as the main source of fresh oxygen.

The ECLSS is very efficient at recovering waste water – not just from the obvious sources, but also from excess moisture in the atmosphere, including evaporated sweat. Nevertheless, there's still some unavoidable loss in the system. To compensate for this, fresh water must be sent up to the ISS every few months, along with food and other supplies. In the course of a year, this may amount to over a tonne of extra water to ensure the six-person crew stays fit and healthy. That simply won't be an option on a trip to Mars – and nor will fresh supplies of food or anything else.

Currently astronauts on board the ISS can grow their own lettuce – and eat it – but only in small quantities. Even then, the raw materials – seeds and nutrients – need to be supplied from Earth. For a crew heading to the Red Planet, there's no alternative but to carry along everything they're going need with them right from the start.

Roadmap to Mars

After decades of vacillation over its Martian ambitions, NASA finally published a 'roadmap to Mars' in 2015. The name is taken from the well-established practice of technology roadmapping. Originating in the electronics industry of the 1970s, this has since become a standard tool in engineering planning. Technology roadmaps come in a variety of forms, but the general idea is to provide a clear visual representation of the way forward. Roadmaps may reflect either a goal-driven or an innovation-driven philosophy. 'Roadmap to Mars' certainly sounds like it falls in the first category, just as the Moon programme did in the 1960s – but a closer look reveals a distinct legacy from the innovation-driven Space Shuttle era:

Human Exploration: NASA's Journey to Mars*

Earth reliant *Mission: 6 to 12 months* *Return to Earth: Hours*	**Proving ground** *Mission: 1 to 12 months* *Return to Earth: Days*	**Earth independent** *Mission: 2 to 3 years* *Return to Earth: Months*
• Mastering fundamentals aboard the International Space Station • US companies provide access to low Earth orbit	• Expanding capabilities by visiting an asteroid redirected to a lunar distant retrograde orbit • The next step: travelling beyond low Earth orbit with the Space Launch System rocket and Orion spacecraft	• Developing planetary independence by exploring Mars, its moons and other deep space destinations

In its full form, NASA's roadmap goes into considerable detail – the table here just shows a top-level summary. This version doesn't indicate a timescale, but NASA's aim is to get a human crew to the 'vicinity of Mars' sometime in the 2030s.

The plan is divided into three broad steps. The first is labelled 'Earth Reliant', meaning activities carried out in Earth orbit. That's where NASA has been for the last 40 years, of course … so what does it have to do with getting to Mars? Well, it allows NASA to de-risk key aspects of a long space flight in a comparatively safe environment – especially human factors such as health and ergonomics, as well as communications and other basic enabling technologies.

---

* Text from a NASA graphic which can be found at http://solarsystem.nasa.gov/images/galleries/marsextensibility-ready-4sls.jpg

NASA calls its second step 'Proving Ground'. This involves testing technology beyond the cosily familiar territory of Earth orbit – somewhere closer to the Moon, for example. As Neil Armstrong famously pointed out in 1969, that's a giant leap. The ISS circles the Earth just 400 km above the surface, while the Moon can be as much as 400 *thousand* kilometres away at the most distant point of its orbit. That's far enough from home to provide a reasonable facsimile of the deep space environment, but close enough that astronauts can get back to safety in just a few days if an emergency arises.

This phase of the plan is pencilled in for the 2020s. It requires two important new pieces of hardware – a heavy-lift launcher in the Saturn V class, together with a 21st-century upgrade of the Apollo spacecraft. The former has been given the unimaginative name of the Space Launch System (SLS), while the latter is called Orion (although it has no connection with the bomb-powered 1950s nuclear spacecraft of the same name).

Orion's crew module will be larger than the Apollo command module, with around 50 per cent more interior volume, and constructed from more modern materials. While Apollo was designed to carry three astronauts, Orion will have space for up to six seats – though not all of them would need to be occupied on every flight. As with Apollo, Orion will have a large service module to provide power, life support and propulsion. The current plan is for this module to be built by the European Airbus consortium, to a design produced by NASA's international partner ESA. On this plan, the service module will simply be a modified version of ESA's Automated Transfer Vehicle (ATV), which was used as a supply ship for the ISS.

The first flight of the Orion crew module took place in December 2014. There were no astronauts on board, and it was attached to a non-functional mock-up of the service module. The spacecraft spent four hours in orbit before returning to Earth and splashing down in the Pacific. The test was a success, with all systems remaining nominal through-out the flight ('nominal' is NASA jargon for 'according to plan').

That first test didn't require the SLS – which was for-tunate, because it hasn't been built yet. It's still quite a way in the future. The version that's intended to rival the Saturn V is called 'SLS Block 2', and it won't fly until the late 2020s. Long before then, however, a smaller 'Block 1' version – with about half the lifting capacity – will be used for preliminary testing. On its first flight, it will be used to send an uncrewed Orion capsule around the Moon and back to Earth. That may happen as early as 2018. If it's successful, then a second test, using an uprated 'Block 1B' launcher, will repeat the same flight profile using a crewed Orion spacecraft. But that won't happen just three or four months after the first test, as might be expected from the speed at which the Apollo programme progressed – it'll be more like three or four *years* later.

NASA's graphic includes a rather cryptic item in the 'Proving Ground' segment: 'Visiting an asteroid redirected to a lunar distant retrograde orbit'. Sorry, what? Isn't this supposed to be a roadmap to Mars? Well, yes it is – but it's a very cautious, step-by-step roadmap. Everything will become clear in due course.

The asteroid-redirect mission is an idea NASA has been kicking around since the 1990s. At the time of writing its future is in doubt, with proposals for an alternative initiative (see page 104) being circulated owing to pressure from the Trump administration, but in its most recent form the plan involves employing a robot probe to retrieve a boulder-sized rock from an asteroid and bring it closer to Earth. The target object wouldn't be in the main asteroid belt, which is out beyond the orbit of Mars, but would be one of the nearer asteroids whose orbit crosses that of Earth (these are the ones that occasionally smash into the planet with catastrophic results, so we have a vested interest in learning as much as we can about them). Although such asteroids may have brief encounters with Earth, they spend most of their time at much greater distances, so there's a logic to transporting a chunk of one to a location where it can be studied more easily. The specific orbit mentioned – a 'distant retrograde orbit' around the Moon – was chosen because of its long-term stability, so there's no danger of the asteroid eventually falling to Earth. NASA's proposal suggests the asteroid fragment would be in place by 2025, allowing an Orion crew to visit it soon after that.

Even when the asteroid redirect mission is spelled out in detail, the reason for it – and its relevance to Mars – is still far from obvious; hence the misgivings of President Trump and other American politicians who have seen it as a 'time-wasting distraction' in the words of one member of Congress. But that's not the way the engineers at NASA see it. They may go about things in a slow and roundabout way, but they always have a reason for what they do.

The purpose of the asteroid mission becomes clearer when we look at the third phase of NASA's plan, labelled 'Earth Independent'. That sounds promising – does it finally mean we're going to land on Mars? Maybe eventually, but not to start with. The initial aim of this phase is simply to get to *the vicinity of Mars*. Specifically, what NASA has in mind is exploration of the Martian moons Phobos and Deimos. The fact that they're so tiny, with a very low surface gravity, makes them a much simpler objective in space flight terms than a landing on the surface of Mars.

A visit to Phobos or Deimos wouldn't be a pointless exercise. It would allow up-close observation of the Martian surface, and more effective operation of rovers due to the absence of any significant radio delay. It might even permit some degree of ISRU, because the moons may contain water – in the form of ice or hydrated minerals – and other useful constituents. But there's nothing particularly special about Phobos and Deimos – as stated previously, they're probably just ordinary asteroids that were captured by Mars's gravity. So the asteroid redirect mission would allow NASA to test all the necessary technologies and procedures much closer to Earth.

The big advantage of the captured asteroid is that it would only be a few days' journey from Earth, so astronauts visiting it could manage with just the Orion spacecraft. The Mars mission, on the other hand, would entail journeys of well over a year, with no quick way home in case of emergencies. That's why this phase of the plan is called 'Earth Independent'. It requires a much larger spacecraft – the Deep Space Habitat (DSH) mentioned in Chapter 4. Unlike Orion, which is already a finished design, the DSH is just a name and a few sketches on paper. No one knows exactly what the final version will look like, or even who will build it. But the most likely configuration is a grouping of ISS-derived modules for the crew, together with a newly designed propulsion unit. This might be an electrically powered ion drive or, more likely, a conventional rocket engine.

Deep Space Gateway

Just as this book was going to press, in May 2017, NASA presented a somewhat revised Mars programme at a 'Humans to Mars' summit in Washington DC. Similar in broad outline to the 2015 roadmap, it replaces the asteroid-redirect mission with a Moon-orbiting space station, dubbed the 'Deep Space Gateway', which would fulfil a similar risk-reduction role. Among other things, it would facilitate a year-long 'Mars simulation mission' – pencilled in for 2017 – using the same hardware as a real Mars mission but taking place in the relative safety of the Earth–Moon system.

Overall, NASA's plan for Mars looks feasible. The necessary technologies already exist – or have done in the past – from Apollo and the Saturn V to Curiosity and the ISS. It's just a matter of putting the pieces together and making sure they work the way they're supposed to. So it comes as a shock to look at NASA's timescales and see how long that's going to take: several years to the first crewed flight of Orion, several years de-risking technology in lunar orbit, several more years before the first Mars mission – and then only to its moons, not the surface of the planet.

It's no wonder some people think they can do it faster.

PRIVATE ENTERPRISE 6

Commercial space flight

> 'It looks like we've caught a Dragon.'

Manipulating the external robotic arm of the ISS on 10 April 2016, British astronaut Tim Peake had indeed 'caught a Dragon' – a Dragon resupply capsule, designed, built and operated by the private firm SpaceX. It was the seventh SpaceX Dragon to make a successful docking with the space station, and a triumph for private enterprise in more ways than one. Packed up on board was a novel piece of technology called the 'Bigelow Expandable Activity Module', or BEAM, produced by the Bigelow Aerospace company. This was nothing less than an inflatable extension for the ISS, which the astronauts successfully installed the following month. And Dragon wasn't alone, either. Sitting at the next docking berth was another privately operated resupply vessel – an Orbital ATK Cygnus capsule.

**The SpaceX Dragon capsule arriving
at the ISS on 10 April 2016.**

(NASA image)

The private sector is no newcomer to space travel. Apollo was built by a private company, North American Aviation, while the Saturn V was a joint effort by North American and Boeing. Apollo and Saturn V were NASA designs, however, and the companies that built them did so under contract to NASA. The novelty that has emerged in recent years is the concept of a spacecraft that is 100 per cent private enterprise – all the way through from design and construction to testing and operation.

The funding for a space flight may still come ultimately from the government, but the difference is that it's now paying for a service – just as it might pay for a courier service – rather than buying a piece of hardware. This new business model saw its first practical application in NASA's

Commercial Resupply Services (CRS) – a series of contracts for the delivery of cargo and supplies to the ISS via commercially operated spacecraft. The contracts are split between two suppliers: Elon Musk's SpaceX, using the Dragon spacecraft and Falcon 9 launch rocket, and Orbital ATK with the Cygnus capsule and Antares launcher.

A key aspect of this approach is that it places the financial risk on the commercial supplier rather than the 'customer'. Although NASA pays for the CRS missions when they're delivered, all the up-front development costs had to come from private investors. That's a big change from the old way of doing things, but it seems to work – arguably better than anything NASA did on its own in the last 40 years.

It's difficult to believe in hindsight, but it was only in September 2008 that SpaceX achieved its first successful space flight, when a Falcon 1 launcher placed a dummy satellite in orbit. Just over two years later, in December 2010, a larger Falcon 9 rocket put the first Dragon spacecraft into orbit. By October 2012 Dragon had started to earn its living, by taking a load of supplies to the ISS in its first CRS mission. Less than a year later, in September 2013, Orbital ATK did the same thing with one of its Cygnus vehicles.

This is an impressive rate of progress, particularly when compared to the glacial pace at which NASA's intramural Orion project is inching forward. And it's despite the fact that the private companies are battling against a steep learning curve – or a series of failures, to put it more bluntly. Orbital's third CRS mission in October 2014 ended prematurely when its Antares rocket exploded soon after lift-off. SpaceX managed six successful CRS missions before their first failure, when they suffered a similar post-launch explosion in June 2015. The company endured a second setback in September

2016, when another Falcon 9 rocket – not destined for the ISS this time, but intended to put a commercial satellite into orbit – blew up on the launch pad during pre-flight testing. Frustrating as they are, setbacks like these are just par for the course when new rockets are being developed. Back in the 1950s, for example, half of America's first ten satellite launches ended in failure. Seen in this perspective, SpaceX and Orbital ATK are doing pretty well.

Private companies aren't just playing catch-up with NASA. SpaceX in particular is doing something completely new with Dragon – it's the first ISS resupply module that comes home after finishing its job. Its government-built predecessors like the European ATV and Russian Progress were designed to burn up in the Earth's atmosphere, as does Dragon's commercial competitor Cygnus. But Dragon is built like an Apollo command module, with a heat shield and par-achute – and it can splash down in the Pacific just as Apollo did. That means Dragon could quite easily be adapted to human flight – it has a total pressurised volume comparable to NASA's Orion crew module. Elon Musk has joked that, even in its present form, Dragon would allow a stowaway to survive a flight to the ISS and back.

A genuinely human-rated Dragon, called 'Dragon 2', is already in the works. Using the same basic capsule design as the cargo Dragon, and with the same dimensions, it will seat up to seven astronauts – one more than NASA's Orion. Its initial use will be to transport crew members up to the ISS, under a CRS-like arrangement called the Commercial Crew Programme (CCP). The first such flight could take place as early as 2018.

Just like earlier crewed capsules such as Apollo and Soyuz – and the current cargo Dragon – Dragon 2 will enter the Earth's atmosphere using a heat shield and aerodynamic braking to slow it to subsonic speeds. But from that point on, the Dragon 2 design goes a step beyond the parachute-slowed descent of its predecessors. With Dragon 2, the parachute will only be there for emergencies: its normal landing mode will use a downward-pointing retro-rocket to slow it down, all the way from the edge of space to touchdown on solid ground. The spacecraft will land vertically, on four extensible legs, with the pinpoint accuracy of a helicopter.

This sort of gentle, controlled landing not only looks more dignified than an Apollo-style splashdown, but it has a practical benefit too. It means the spacecraft can be reused over and over again for different missions. In the same way, SpaceX's Falcon 9 rocket is designed to be recovered in one piece, ready for reuse. That's a completely different philosophy from NASA launchers, which use up all their fuel during the ascent, and then just fall back into the sea. In contrast, Falcon 9 continues to use its rocket on the way down – for deceleration instead of acceleration – and lands vertically on its tail in a precisely chosen spot. That's a difficult trick to pull off, and SpaceX's first few attempts ended in failure. But the technique was finally perfected in December 2015, and has been repeated several times since.

SpaceX wasn't the first company to carry out the controlled landing of a launch rocket. That honour went to SpaceX's competitor, Blue Origin, run by Amazon founder Jeff Bezos. Just like Falcon 9, Blue Origin's New Shepard rocket is designed to land vertically – a feat it achieved for the first time in November 2015. The same rocket went on to fly, and land, four more times. But New Shepard is less powerful than Falcon 9, and not capable of putting a payload into orbit. Instead it's limited to suborbital flights, like that by which Alan Shepard – after whom Blue Origin's rocket is named – became the first American in space in 1961.

Blue Origin's business model isn't the same as that of SpaceX. Jeff Bezos is targeting a completely new customer for space launch services: the general public. Space tourism is a novel market, but it's one that Blue Origin – and a number of rival companies – can grow gradually, feeling their way forward as they learn what pays off and what doesn't. In effect, the company is following NASA's innovation-driven rather than goal-driven model, with the first few steps simply aimed at establishing the viability of a space tourism business in the first place.

Blue Origin's slow but determined approach is reflected in the company's Latin motto: *gradatim ferociter*, meaning 'step by step, ferociously'. The corporate logo features a pair of tortoises, in a deliberate allusion to Aesop's fable of the tortoise and the hare. In that story, it was the tortoise that eventually won the race – and in the real world, the first living creatures to fly around the Moon were a pair of Russian tortoises, on board Zond 5 in September 1968. When it comes to the private-enterprise space race, the hare is undoubtedly Elon Musk's SpaceX – but Blue Origin may not be far behind.

The reusability that is such a watchword at these companies isn't a new concept – after all, the orbiter component of NASA's Space Shuttle was reusable. But it had to be, because it was so complex and expensive. Coupled with a much simpler design, as in New Shepard and Falcon 9, reusability becomes a route to major cost savings. That makes space travel much more affordable, which will be a key factor in the race to Mars. As Elon Musk put it in June 2015:

If one can figure out how to effectively reuse rockets just like airplanes, the cost of access to space will be reduced by as much as a factor of a hundred. A fully reusable vehicle has never been done before. That really is the fundamental breakthrough needed to revolutionise access to space.

Musk's whole attitude to space is different from NASA's. Where NASA emphasises safety and reliability, Musk emphasises cost-effectiveness. Where NASA takes a slow and cautious approach, Musk gets things done as quickly as possible. That's a reflection of longstanding cultural differences between the public and private sectors, but given that NASA is SpaceX's number one customer, it can lead to friction. An example of this emerged in the wake of the launch pad explosion that SpaceX suffered in September 2016. When the cause of this was eventually tracked down, it turned out to be something that could never have happened with one of NASA's rockets – or with anybody else's except SpaceX's.

One of the unique features of the Falcon 9 launcher is the way it uses very cold liquid oxygen. It's not the liquid oxygen that's unusual – most rocket launchers use it – but the fact that SpaceX cools it to a much lower temperature than everyone else, close to oxygen's freezing point. This 'subcooling', as it's called, is done to increase the power and efficiency of the rocket, but it has the side-effect of making the oxygen much trickier to handle. What seems to have happened in the 2016 accident is that the oxygen was cooled so much that it actually froze. If there had been a crewed Dragon capsule on top of the rocket instead of a communications satellite, the resulting explosion could have been fatal.

That's something that bothered the NASA people a lot. Not only would they never subcool oxygen to that extent, but they would never dream of fuelling a rocket with the crew on board. With SpaceX's approach, however, the oxygen must be loaded at the last minute – which in the case of a crewed mission, would mean *after* the astronauts had boarded and were all set to go. Interestingly, one of the people who spoke out against the practice was former astronaut Tom Stafford – the man who, back in 1969, was so safety conscious that he squashed the suggestion that his Apollo 10 mission should be upgraded from a test flight to a lunar landing. Now chairman of NASA's ISS Advisory Committee, he said that any kind of fuelling is a 'hazardous operation' and that nobody should be near the launch pad while it is in progress.

Making life multiplanetary

SpaceX is happy to provide NASA with Earth-orbital services like CRS and CCP because those are good, solid ways to make money from space travel. But the company's real interests lie further afield – as indicated by its corporate motto, 'Making Life Multiplanetary'. Just what that entails was explained in some detail in a presentation by Elon Musk at the 67th International Astronautical Congress in September 2016. The talk focused on a journey to Mars – an undertaking that Musk clearly believed was not only technically feasible but commercially viable too. He put a strong emphasis on reusability and the efficient use of fuel resources, via refuelling in orbit and ISRU on Mars.

Musk didn't pull his punches when it came to future space hardware, either. He described how SpaceX would develop a brand-new launch vehicle for flights to Mars – and beyond – which would be three or four times as powerful as a Saturn V. Provisionally called the Interplanetary Transport System (ITS), it will probably acquire a catchier name at some point in the future. Although the ITS is just a concept on paper, its massive Raptor engine is already in development; its first test-firing took place the day before Musk's talk. An unusual feature of the Raptor engine is that it uses methane rather than kerosene or hydrogen as a fuel. Part of the reason for this is that methane can be produced more cheaply than hydrogen, while being better suited to vehicle reuse than kerosene. Even more important, methane is by far the easiest fuel to synthesise on Mars – making it the obvious one to standardise on, in Musk's view.

The ITS launch vehicle will employ 42 Raptor engines – and it will need every one of them, because sitting on top will be a proportionately gigantic payload. Musk calls it a 'spaceship' – a term that's been hopelessly over-hyped in the past, but is justified in this case. At 50 metres in length – half the size of a football field – Musk's proposed spaceship would have room for up to a hundred passengers and crew. The whole thing would land on Mars, synthesise some more fuel, and blast off again for Earth. Both the spaceship and its launch vehicle would be fully reusable – over and over again – and that's how Musk believes he can get the cost per tonne down to an affordable level. He suggested that SpaceX flights to Mars might start as soon as the mid-2020s.

Musk's 2016 presentation was aimed more at enthusing wealthy investors than convincing hard-nosed engineers. As for the latter – it's obvious from Musk's record that he knows how to separate fact from fiction, and reality from speculation. SpaceX's near-term plans for Mars look eminently achievable, centring on an uncrewed variant of Dragon 2 called Red Dragon. This would be sent to Mars using SpaceX's new Falcon Heavy rocket – a more powerful version of Falcon 9. The first launch could be as early as the 2018 Hohmann window ... and unlike a mission by NASA or ESA, it won't be preceded by endless hand-wringing over its scientific objectives. The purpose of the flight will be clear and simple: to test equipment and procedures that could be used on a crewed mission at the next Hohmann window, or maybe the one after that.

The beauty of using a Dragon 2 variant is that it's a spacecraft that already knows how to land on a planet. With relatively minor modifications, the same combination of heat shield and propulsive landing that would be used on Earth would work on Mars – as shown in this artist's impression:

Artist's impression of Red Dragon landing on Mars.

Elon Musk isn't the only person thinking about how to get to Mars with SpaceX hardware. In 2013, Robert Zubrin proposed a scaled-down version of his Mars Direct plan using

Falcon rockets and Dragon spacecraft. This would entail three Falcon Heavy launches – one with a crewed Dragon module, and the other two carrying support equipment. The latter would include a MAV, which is the only completely new piece of kit that would be needed. The Dragon capsule has a maximum capacity of seven people, but in Zubrin's proposal there would be just two astronauts on board. Additional space for the journey would be provided by a six-metre-diameter, eight-metre-long inflatable extension – effectively a scaled-up version of the BEAM module that was added to the ISS in May 2016. Constructed from thick layers of flexible fabric and an aluminium frame, this can be packed up tightly for transport to orbit, and then expanded to its full dimensions when pressurised. The extension envisaged by Zubrin would provide each crew member with over a hundred cubic metres of space, which is a comfortable amount for a flight to Mars and back.

Human interest

For its current operations, SpaceX's business model may be novel in the space context, but it's basically a conventional one. They offer a service – putting payloads into Earth orbit – and customers pay for that service. Some of their clients are commercial companies wanting to put private satellites into orbit, but the bulk of SpaceX's income still comes from the US government – and ultimately the taxpayer – via NASA contracts such as those for CRS and CCP.

That's not going to be the case with human flights to Mars. They aren't going to offer an immediate return on investment in the way that orbital operations do. Elon

Musk's vision for the Red Planet is different from NASA's. Their focus is scientific, his is human-centric. That calls for a completely new approach to the funding of space missions.

Just how this will be achieved remains to be seen. In his much-publicised Mars presentation in September 2016, Musk was clearly seeking the support of super-rich private individuals – by offering them not a financial return but a stake in the future. Part of this was an appeal to an altruistic, philanthropic spirit: the opportunity to help shape the future of the human species, by enabling it to expand beyond Earth … and maybe even prevent its total extinction as and when the next global catastrophe strikes our own planet. On a lighter note, why wouldn't someone who could afford the ticket price want to go to Mars, simply for the pleasure and excitement of the trip? Here Musk was addressing the sort of person who might otherwise splurge their hard-earned cash on a 10,000-tonne superyacht or a small tropical island.

Seen in this context, the sheer human interest of a Mars mission becomes a driving factor. When Elon Musk talks about Mars, the engineering is still there in the background but the main spotlight is on the human payload – and it's no longer just a 'crew', but 'passengers' as well. Other Mars enthusiasts go even further, focusing exclusively on the human aspects of a journey to Mars at the expense of the (to many people dull and boring) technological ones.

Issues such as these become critical in the context of a self-funded journey to Mars. As science fiction author Gregory Benford showed in his 1999 novel *The Martian Race*, there's no fundamental incompatibility between undertaking a voyage of exploration and exploiting its human-interest angle to make money. He cites the specific example of the Antarctic explorer Ernest Shackleton:

For Shackleton, self-promotion had been essential all the way. He had paid all his expenses with media tie-ins, one way or another: auctioning off news and picture rights before he left, taking special postage stamps along to be franked at the south pole. After he made it, his best-seller had nine translations. He spruced up his expedition ship into a museum and charged admission. With a lecture tour and phonograph record, a first film of the Antarctic and countless newspaper interviews, he made his way into history – and prosperity.

A real-world counterpart to Benford's fictional enterprise is the Mars One initiative. This originated in 2012 as the brainchild of Bas Lansdorp, a young entrepreneur from the Netherlands. That country, of course, was the birthplace of the archetypal reality TV show, *Big Brother* – and it's reality TV that provides the financial engine for Lansdorp's proposal. Put simply, Mars One would be funded by modelling the whole thing on *Big Brother*. At first sight, the idea makes sense – there's no doubt that big money can be made from a successful TV show. But is it an approach that's compatible with the practicalities of spaceflight? Reality TV thrives on conflict situations – often artificially contrived by the producers to increase ratings – and conflict situations are exactly what a space mission doesn't need. As mentioned in Chapter 4, the selection process for astronauts is pretty much the opposite of that for reality show contestants. If Neil Armstrong had auditioned for *Big Brother*, the producers would have written 'boring' on their clipboards and sent him packing.

It's ironic that the immediate post-Apollo years saw the first *Star Wars* movies, and the consequent rise of the

Hollywood sci-fi blockbuster. From that point on, fictional portrayals of space travel – which had previously only appealed to a niche audience – were guaranteed to capture the imagination of the public at large. At the same time, there's been an opposite trend with real-life space missions, which rarely attract attention outside a small subgroup of enthusiasts. In recent years, the closest a space feat came to causing genuine public excitement was when ESA's Rosetta mission landed a small probe on comet 67P in 2014. Even then, amid all the buzz about Rosetta's scientific and technical achievements, there was a human-interest subplot worthy of *Big Brother* – centring on a politically incorrect shirt that one of the project scientists chose to wear at a press briefing.

With regard to human space flight, people often point to the Apollo 11 landing as a time when 'the whole world was watching' … but that was literally only for a few days. On a Mars mission, interest would need to be maintained week after week for several years. That's simply not going to happen if everything remains 'nominal', as NASA would put it. You can imagine producers desperately hoping for an Apollo 13-style crisis.

In keeping with its human-interest focus, Mars One's activities to date have dealt with things like choosing the crew and designing the Martian habitat. The nuts-and-bolts technical details of how they're going to get from Earth to Mars are still very vague. The plan seems to rely heavily on other organisations – SpaceX in particular – doing the hard work on technology development, allowing Mars One to buy the necessary kit 'off the shelf' when it's ready for use.

Most of what's been written about Mars One – by everyone except Lansdorp's own team – has been overwhelmingly negative. An independent study by the Massachusetts

Institute of Technology came to the blunt conclusion that 'the Mars One mission plan, as publicly described, is not feasible'. Even one of the project's avowed supporters, the Nobel Prize-winning physicist Gerard 't Hooft, suggested that both Lansdorp's timescales and his costings were too optimistic by a factor of ten.

The biggest flaw in the Mars One proposal may not lie in its business model or its technical feasibility, but simply in the fact that it focuses on Mars in the first place. If Lansdorp's 'unique selling point' is the concept of reality TV in space, why does it have to be on Mars? Why make things difficult by attacking a whole series of problems that far more experienced organisations haven't solved yet? It would be much simpler to put the contestants in a Skylab-style space station in Earth orbit. That would establish the viability of Lansdorp's business model, with a fraction of the risk and a fraction of the up-front investment. Maybe the series would be a success, and maybe it wouldn't. If it did end up making a profit, that would be a good time to start thinking about a follow-up series on Mars.

Around the same time as Mars One, a somewhat less ambitious proposal came from an American businessman named Dennis Tito, under the auspices of his Inspiration Mars Foundation. Tito's name has already gone down in space history, as the world's first space tourist. Back in 2001, he bought a seat on a Soyuz flight to the ISS, and spent a week there. Just as he was happy to pay for that trip, he believes a privately financed flight to Mars is a viable proposition.

As with Mars One, Tito's plan is strong on human interest and weak on science. Nevertheless, he's worked through the technical details more carefully than Lansdorp, and the

Inspiration Mars proposal looks to be feasible – at least in principle. It would see just two people on a fly-by mission around the Red Planet, followed by an immediate return to Earth – there would be no tricky manoeuvring to go into orbit, let alone a landing on the surface. The crew would consist of a married couple, on the grounds that they would find it easiest to get on in each other's sole company for the 500 days the flight would last.

On the face of it, Tito's proposal is a perfectly reasonable first step – though its main achievement would be a symbolic one, allowing the human species to collectively say 'yes, we've been to Mars'. But the flight would have practical benefits too, by testing out some of the hardware and systems needed for later, more ambitious missions. The problem is that, just like Mars One, Inspiration Mars will only work if it can buy all the necessary hardware from SpaceX. And therein lies the catch, because Tito's mission profile doesn't fit in with SpaceX's own plans. Elon Musk isn't interested in flying round Mars and coming straight back – he wants to see humans living there permanently.

LIVING ON MARS

7

Colonisation

> I think there are really two fundamental paths. History is going to bifurcate along two directions. One path is we stay on Earth forever, and then there will be some eventual extinction event – I don't have an immediate doomsday prophecy ... just that there will be some doomsday event. The alternative is to become a space-faring civilisation and a multi-planet species.

When Elon Musk spoke at the International Astronautical Congress in September 2016, he made Mars colonisation sound like a necessity, not a luxury. Confining ourselves to just one planet is like putting all our eggs in one basket. Musk's reference to 'some doomsday event' isn't idle scaremongering – it's the plain truth. The Earth's long record of species extinctions shows that, given enough time, planet-wide catastrophes really do happen. A foothold on Mars would mean humanity would always have that second chance.

Musk's Martian colony would be set up using SpaceX's huge, reusable ITS – the Interplanetary Transport System – carrying a hundred people at a time to the Red Planet. This wouldn't happen just once or twice, but on a continuing schedule, with at least one flight at every Hohmann launch window. When Musk talks about 'making life multiplanetary', he means it. Some passengers on the ITS would be short-term tourists, but others would stay and work on Mars, helping to develop a self-sustaining infrastructure.

Living on another planet won't be easy, and unlike some Mars enthusiasts Musk doesn't gloss over the difficulties. Early settlers will be highly reliant on supplies from Earth. There are useful in situ resources on Mars – that's one of the reasons for picking it out, rather than, say, the Moon – but all those resources require specialised equipment to get at them. Water can be extracted from subsurface ice, and oxygen from atmospheric carbon dioxide. Growing plants on Mars will help with the latter process, and of course provide a source of food at the same time. Electrical power can be drawn from the Sun, courtesy of large fields of solar panels. But all these things require hardware that has to be shipped from Earth. So Musk's plans include a fleet of large cargo ships in addition to the human-carrying ones.

Musk's long-term vision isn't unique to him. Similar sentiments have been voiced by none other than NASA administrator and former Space Shuttle astronaut Charles Bolden. In April 2014, in what almost sounds like a plug for SpaceX, he said:

> If this species is to survive indefinitely we need to become a multi-planet species, we need to go to Mars, and Mars is a stepping stone to other solar systems.

Significantly, Bolden spoke these words in defence of NASA's gradual 'Earth Reliant–Proving Ground–Earth Independent' route to Mars discussed in Chapter 5. The point he was emphasising is that landing on Mars is not an end in itself, but just one step in a long, continuous process. This is an important difference from the Apollo-style Mars mission that President Nixon vetoed in the 1970s. If those plans had gone ahead, NASA astronauts might have landed on Mars in the 1980s – but like their predecessors on the Moon, there's a danger they would have simply planted a flag, taken a few rock samples and left it at that. In fact, that's exactly how Stephen Baxter portrays events in his alternate-history novel *Voyage*. Seen in that light, maybe a slower start is justified, if it ultimately leads to a self-sustaining colony on the Red Planet.

Both Musk's and Bolden's plans for Mars colonisation are feasible and realistic, even if they're long-term. Other people have proposed a different approach, which sees a Mars colony as the *first step* in the process, rather than an eventual goal. This is the 'Mars to Stay' philosophy, which has been promoted by former astronaut Buzz Aldrin and several others. Their reasoning starts from the unarguable fact that a one-way trip to Mars is easier than a two-way journey. So, they say, why not crew a spaceship with permanent settlers instead of transient explorers? The idea is that this would be much cheaper, because it would not need a MAV or return craft. But this ignores the vast amounts of hardware and supplies that would have to be shipped from Earth to keep the colony going until it became self-sufficient. Sadly, it's difficult to see Mars to Stay as anything but an exercise in false economy.

A slower, more cautious route to building up a

sustainable presence on Mars makes a lot more sense. It might take decades, but eventually the nascent colony could obtain all its air, water, fuel and building materials from the local environment. The colonists could grow all their own food – possibly planted in native Martian soil, with the addition of appropriate nutrients, inside huge greenhouses. The planet certainly receives enough sunlight to make this a practical proposition. It's further from the Sun than the Earth is, but on Mars there's no such thing as a really cloudy day – just the occasional dust storm. There's another, more serious, problem though – and that's radiation.

As mentioned in Chapter 4, the principal radiation hazard in space comes from the fast-moving charged particles making up the solar wind and cosmic rays. Here on Earth we're protected from such radiation by our planet's magnetic field, which deflects the path of charged particles away from us. But Mars doesn't have a magnetic field, so cosmic rays and the solar wind beat down unimpeded on its surface. That creates a dilemma for any would-be Martian horticulturalist: plants need sunlight to thrive, yet by exposing them to sunlight in a greenhouse they're also being exposed to harmful radiation.

Fortunately there's another way to cultivate crops that may be more suitable for a Martian colony. Hydroponics is the practice of growing plants in nutrient-saturated water, rather than soil, under arrays of artificial lamps instead of sunlight. On Earth, it's often associated with the surreptitious cultivation of *Cannabis indica* – but it's also a long-time favourite of science fiction authors, who realised early on that it would be the easiest way to grow food on a spaceship. That's perfectly true, and the system the ISS astronauts use to grow lettuce is a hydroponic one. But it's also the obvious thing to use on Mars, because it allows crops to be raised in carefully controlled conditions in an environment that's well-shielded from radiation. The process would still use sunlight – but only indirectly, via solar panels powering arrays of LED lights.

What's true for plants is, of course, true for human colonists too: they're going to need trustworthy day-to-day protection against cosmic radiation. For the earliest missions, surface habitats will have to be brought all the way from Earth – which means they will need to be lightweight. Most likely they will be expandable prefabricated modules – like the BEAM extension on the ISS, mentioned in the previous chapter. On its own, such a habitat would provide adequate protection for a short-stay mission of a month or so, but not for a significantly longer stay. In that case, the same basic habitat could be used, but it would need to be shielded from radiation by partially burying it in the Martian soil. Another option, if subsurface ice is as abundant as many people believe, is a kind of Martian igloo – an inflatable dome surrounded by a thick shell of ice.

For anything approaching a permanent colony, an inflatable habitat would be totally inadequate. Since it's impractical to transport anything more substantial from Earth, the only alternative is to make something in situ, using locally available materials. This might be achieved using time-honoured construction techniques – building houses, solid enough to be pressurised, from the ground up – or the colonists might go for a more high-tech alternative in the form of 3D printing. In 2015 NASA ran a competition 'to develop architectural concepts that take advantage of the unique capabilities 3D printing offers to imagine what habitats on Mars might look like using this technology and in-situ resources'. One of the entries is shown in the following illustration.

One of the entries in NASA's
'3D Printed Habitat Challenge'.
(NASA image)

Given enough time, energy and ingenuity, human colonists could make Mars a survivable place to live. For long-term viability, though, the colony needs to do more than just survive. It needs to thrive, both economically and culturally. How would it do that?

Martian sustainability

The fact that Mars is 'in space' – and that its first-generation colonists would all be space travellers – suggests that perhaps they could earn their living from space. There's a potentially vast source of wealth nearby, in the form of the main asteroid belt. This is significantly closer to Mars than it is to Earth – not just in the conventional sense of kilometres, but also in the rocket-science sense of delta-v. As seen in Chapter 2, this is the critical thing when it comes to the practicalities of space flight – and the delta-v needed to get to the heart of the asteroid belt is halved if the starting point is Mars rather than the Earth.

The asteroids are made from the same basic elements as the Earth itself – so that includes socially and industrially valuable metals like gold, platinum, rhodium and palladium. The small size and weak gravity of an asteroid means that finding and extracting such minerals is – in principle at least – much simpler than on Earth. 'Asteroid mining' is a cliché of science fiction, and many people would like to see it become a reality in the not-too-distant future. In 2012, a private company called Planetary Resources Inc. was set up to explore just that possibility. Whether the revenue amassed by such a venture would ever be sufficient to cover the costs of transportation remains to be seen, however.

Perhaps surprisingly, one person who is sceptical about the idea of a Mars colony engaging in asteroid mining is Elon Musk. He suggested that a more practical export from the Red Planet would be 'something that can be transported with photons as opposed to atoms'. All material objects are made up of atoms, which give them mass – and mass is what pushes up the cost when it comes to transporting things across space. On the other hand, radio signals – like any other electromagnetic waves – travel from A to B in the form of massless photons. That's a lot cheaper, and a lot faster – but it can't transport physical commodities, only information. That's what Musk thinks future Martian colonists should concentrate on exporting back to Earth.

Long-time Mars advocate Robert Zubrin agrees with Musk, suggesting that one major 'export' could be a stream of new technological inventions, in the form of patents. There's some sense to this – it's a fact, for example, that the companies with the highest annual turnover per employee tend to be ones that own a large number of registered patents. They earn the bulk of their income through the licensing of 'intellectual property' to other companies, which then do the actual manufacturing and selling.

The idea that a Martian colony would be a hotbed of scientific innovation comes across in Arthur C. Clarke's 1951 novel, *The Sands of Mars*. It's easy to understand why Clarke saw things that way. At the time he was writing, few people had any interest in space travel, and the ones who did tended to be scientifically literate individuals with above-average intelligence and creativity. Clarke's mistake was to assume this would always be the case, even after space travel became commonplace. History suggests otherwise. At exactly the same time – the mid-point of the 20th century – new technologies like transistorised electronics and digital computers were beginning to make an appearance in physics laboratories around the world. As far as the public was concerned, only the geekiest of science enthusiasts would have been aware of such things – even fewer than knew about the possibilities of space travel. But does that mean that science nerds would be the only people to take up new digital and electronic technologies? That's certainly not the way things turned out.

It's very difficult for us here on Earth, now, to anticipate how a future Martian colony might work out in the long term. Ultimately it will be down to the colonists themselves to create their own way forward. That's the idea behind Isaac Asimov's 1952 short story, 'The Martian Way'. This depicts a time when the colony has begun to think of itself as a distinct society, with its own priorities and values, while still frustratingly dependent on Earth. Meanwhile on the home planet, the novelty of having colonists on Mars has begun to wear off. Some politicians are even calling for the whole project to be dropped, because it's nothing but a drain on Earth's economy.

In Asimov's story, the colonists' biggest problem is water. They have to import most of it from Earth, which is clearly unsustainable in the long term. Their solution – 'the Martian Way' of the story's title – is to 'mine' large chunks of ice from Saturn's rings. As it happens, water probably won't be a major issue for real Mars colonists, because the planet seems to have plenty of ice of its own. This is a rare example of 20th-century science fiction being overly pessimistic – rather than optimistic – about conditions on the Red Planet.

A similar political situation was portrayed by Arthur C. Clarke in *The Sands of Mars*. In that novel, Clarke put the following words in the mouth of the Mars colony's Chief Executive (a man named Hadfield, interestingly enough – in unconscious anticipation of one of the best-known astronauts of the early 21st century):

We're at war with Mars and all the forces it can bring against us – cold, lack of water, lack of air. And we're at war with Earth. It's a paper war, true, but it's got its victories and defeats. I'm fighting a campaign at the end of a supply line that's never less than fifty million kilometres long. The most urgent goods take at least five months to reach me – and I only get them if Earth decides I can't manage any other way. I suppose you realise what I'm fighting for – my primary objective, that is? It's self-sufficiency.

In *The Sands of Mars*, the ultimate obstacle to self-sufficiency isn't as straightforward as a lack of water – it's a basic lack of freedom. No matter how much water, air and food the colonists have, the essential hostility of the Martian environment means everything must be confined inside pressurised domes. Why can't they be free to roam the whole planet, like we do on Earth? In Clarke's novel, the colonists come up with a solution using their scientific ingenuity – which as mentioned earlier, is portrayed as one of their great strengths.

At the time the book was written, Mars was believed to have a much thicker atmosphere than it actually does. The air would still have been unbreathable, due to lack of oxygen, and poisonous, because of the high levels of carbon dioxide. Both problems could be put right – just as they were on Earth half a billion years ago – if there was plenty of vegetation around to convert carbon dioxide to oxygen. But that would require more warmth and sunlight than Mars gets. Clarke's solution was to initiate a self-sustaining 'meson resonance reaction' in the core of Mars's inner moon, Phobos. This turns the little ex-asteroid into a miniature Sun that would continue to burn for a thousand years or so, providing the much-needed heat and light for plant life to thrive on the Red Planet.

Terraforming

The process Clarke describes in *The Sands of Mars* is an example of terraforming – the hypothetical practice of making an alien planet more Earthlike. It's an idea that science fiction writers were kicking around long before real scientists got hold of it, and as far as the latter are concerned it's still a highly speculative topic. Yet the basic idea – that Mars could be made more habitable by heating it up – is a valid one. Some terraforming proponents, such as Stephen Petranek, even believe it would solve the thin atmosphere problem, by evaporating the frozen carbon dioxide in the Martian polar caps. There's a positive feedback loop here, because making Mars warmer thickens the atmosphere, and thickening the atmosphere makes Mars warmer. It all comes down to the greenhouse effect.

The Earth and Moon are the same distance from the Sun, but the Moon gets much colder than the Earth at night. That's mainly because Earth has a greenhouse effect – due to its atmosphere – and the Moon doesn't. Some gases in the Earth's atmosphere – carbon dioxide in particular – hold in heat like glass holds heat in a greenhouse. The term 'greenhouse effect' is often used in a negative way, but that's because too much carbon dioxide makes the Earth too hot. Without any kind of greenhouse effect, it would be much too cold – and that's basically the situation on Mars today.

Petranek's suggestion is that by artificially heating the Martian poles, more carbon dioxide would be released into the atmosphere, a greenhouse effect would be triggered, and the planet would become more Earth-like. But how do you start the process? Arthur C. Clarke's 'meson resonance reaction' won't work, for the simple reason that it's a fictional invention. But there are other ways of achieving the same thing. Petranek suggests providing the heat with giant orbiting mirrors, or by crashing asteroids into the planet to create nuclear-scale explosions, or by seeding it with genetically engineered microorganisms capable of metabolising carbon dioxide.

Although enthusiasts like Petranek make terraforming sound easy, it might be much harder to pull off in practice. If there's one thing we've learned from our own planet, it's that a global ecosystem is an enormously complex, self-interacting mechanism, in which small changes can have disproportionately large consequences. Earth's long history of climate change, with alternating ice ages and much warmer interglacials, has seen oxygen levels fluctuating by a factor of two and carbon dioxide by a factor of ten. It's a notoriously controversial subject, of course – and there's no reason to suppose another planet's atmosphere will be any easier to get to grips with than our own.

In any case, terraforming is based on the assumption that future Mars colonists will want to live in an environment that's as close as possible to Earth's. That's just projecting our own worldview on to future generations. In the spirit of Asimov's 'The Martian Way', the colonists – the second and third and nth generations – will decide for themselves how they want to live. Why should they want to go outside without a pressure suit on? Professional divers don't whinge about the equipment they have to wear – they just get on with the job, or find another one. It will be the same on Mars. Most people will spend all their time 'indoors', and think nothing of it. The 1990 film *Total Recall* gives a vivid picture of what a well-established Martian colony might really be like. It's a long way from a utopia, but at least it's sustainable. Only the first generation of colonists can afford to be idealists; their descendants will have to be realists.

But that lies in the distant future. Anything we write or think about such times will be pure speculation, and probably wrong. One thing is certain, though: long before then, someone will have to take that first step to Mars. Who will it be, and when will it happen?

THE NEW SPACE RACE 8

The contenders

Back in 1999, the science fiction author Gregory Benford wrote a novel called *The Martian Race*, in which he portrayed the eponymous race as playing out in 2018. That time frame may have been too optimistic, but his choice of competitors still looks believable today. Eschewing the government-versus-government rivalry of the Moon race, Benford pitted a government-run team against a privately funded one. On one side was a consortium of private corporations, headed by a charismatic billionaire. On the other side, not NASA or the Russians, but a joint venture by the Chinese government and the European Space Agency. The situation could almost be taken from today's headlines:

- 'The great billionaire space race' (*The Week*, 17 September 2016)
- 'The race to Mars: here's how SpaceX ranks against the competition' (*The Verge*, 30 September 2016)

- 'SpaceX, Blue Origin, NASA compete to put humans on Mars' (*Digital Trends*, 10 October 2016)
- 'Elon Musk versus NASA versus China: billionaires flex financial muscle in new-age space race' (*Daily Express*, 2 January 2017)

At the time of writing (2017), there are only two countries in the world with the operational capability to launch humans into space – and the United States of America isn't one of them. Russia is the obvious one; the less obvious one is China. The first Chinese astronaut was placed in orbit by a Shenzhou spacecraft – similar in design to the Russian Soyuz – in October 2003. Since then, five more crewed Shenzhou missions have flown – one each in 2005, 2008, 2012, 2013 and 2016. Such a leisurely schedule suggests that the Chinese government is less than totally committed to the idea of space exploration. Given the country's other priorities, the situation may be analogous to that in the Soviet Union in the 1960s – when the government was happy to pay lip-service to space, but didn't necessarily follow through with hard cash. That attitude will have to change if China – with or without European collaboration – is to win the race to Mars.

When Benford wrote *The Martian Race*, there was no such thing as a non-governmental space project. But just four years later, the ill-fated Beagle 2 reached Mars with over half its funding coming from the private sector. That showed it could be done, and it has inspired a whole new generation of space entrepreneurs – some with more robust plans than others.

At the less feasible end of the spectrum, there are initiatives like Bas Lansdorp's Mars One and Dennis Tito's

Inspiration Mars, both mentioned in Chapter 6. These proposals make sense up to a point, but they lack key elements of technical detail – particularly regarding the actual space hardware that will be used. That puts these organisations' credibility in question, and sets up a potentially self-destructive vicious circle. The lack of a coherent technical plan leads to criticism in the media, which in turn scares off potential investors. Without significant up-front investment, the companies can't afford the basic technology demonstration and testing needed to silence their detractors.

It's a mistake, in any case, for a recently formed start-up venture to set its sights on an immediate human flight to Mars, given that 50 years of history shows that's a very difficult thing to achieve. Missions to the Red Planet have a depressing habit of failing, even when they're mounted by experienced teams of engineers. In October 2016, ESA made its second attempt – after Beagle 2 – to put a robot lander on Mars … and failed for a second time. Why should a private company, with no prior experience of putting anything into space, imagine it could do better with a crewed vehicle?

The biggest problem with the Mars One approach is not that it's likely to fail, but that by failing it will cast a huge shadow over the whole concept of privately funded space travel. That would be a tragedy, because other people are tackling the same potential market in a much more realistic way. Mention has already been made of Jeff Bezos's Blue Origin company, with its tortoise-based corporate logo. Blue Origin's business model looks far stronger than that of Mars One. While Lansdorp is counting on things like media deals and advertising to raise money, Bezos is planning to offer a tangible service in the form of space tourism. Although Bezos has talked about Mars as a possible future destination, it's not the company's current focus, so it seems unlikely that Blue Origin's tortoise will emerge victorious in the race to the Red Planet – unless everyone else drops out first. In this race, Elon Musk's company SpaceX must be considered the odds-on favourite.

That wouldn't be the case if SpaceX's hopes rested solely on the ITS – Musk's vast Interplanetary Transport System, capable of carrying 450 tonnes of passengers or cargo straight to the surface of Mars. That's great as a long-term vision to inspire potential investors, and it might see preliminary development and testing in the 2020s. But the first SpaceX vehicle on Mars will be much smaller, and based entirely on technology that already exists in the here and now: the Red Dragon capsule, launched by a Falcon Heavy rocket. For preliminary testing Red Dragon will be uncrewed, but flights with humans on board should follow not long after.

Apart from SpaceX, there's really only one contender in the race – and that's NASA. Until very recently, everyone would have seen the NASA team as the obvious, out-and-out favourites. After all, they won the race to the Moon in the 1960s, and they have more experience than anyone else when it comes to getting robotic probes to the Red Planet. But NASA does things slowly, and SpaceX does things fast.

NASA's Orion spacecraft doesn't look all that different from SpaceX's Dragon, but it's taking a lot longer to develop. Even during the few months that this book has been in preparation, NASA's publicly announced plans and timescales have shifted several times. All in all, NASA's perception of the urgency of the race to Mars is altogether more laid back than SpaceX's. The following illustration, showing approximate timelines based on published announcements by two organisations, makes that perfectly clear.

Of course, SpaceX has a good motive for exaggerating the rate at which it expects to make progress, since people are much less likely to invest money in the company if they think they'll have to wait a decade before seeing any tangible results. Politicians can be impatient too, so NASA has a similar problem – but to a lesser extent, and its timescales are probably more realistic. But the chart gives SpaceX a lead of more than ten years – so even if this is out by a factor of two, they'll still win comfortably.

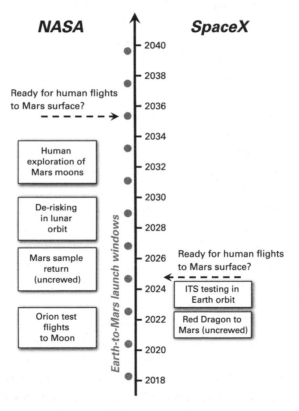

Comparison of NASA and SpaceX plans for Mars.

A wild card

In Andy Weir's novel *The Martian*, NASA is portrayed as winning the race to Mars in 2031 (the date isn't mentioned explicitly in the text, but you can work it out if you read between the lines). The technology employed in the novel is feasible in principle, but significantly more exotic than anything in NASA's current plans. In particular, the use of a nuclear-powered VASIMR propulsion system seems an unlikely development in that timescale. It's not that such a system is too far-fetched from a technical point of view – it isn't – but simply that it would require too many years of development and testing to achieve the necessary levels of reliability and safety certification. Those are the sort of real-world issues that science fiction writers can ignore, but NASA can't.

Although *The Martian* glosses over the time-consuming realities of managing a large engineering project – not to mention the tight-fisted realities of government fiscal policy – it's a different matter when it comes to the even more basic realities of the universe. Weir makes a conscious effort to comply with the laws of physics – from Hohmann orbits and Tsiolkovsky's rocket equation to the conservation of momentum. Science fiction authors aren't usually so fastidious about such things. It would be so much easier if you could travel from Earth to Mars in the same way you travel from London to New York – going essentially in a straight line, leaving as and when it's convenient, using a vehicle with a payload fraction closer to 50 per cent than 3 or 4 per cent of the total launch mass. Countless Hollywood movies give the impression that, at some point in the future, people will be able to do exactly that. But is that likely?

To get from A to B as quickly as possible calls for the expenditure of a lot of energy. However, as seen in Chapter 2 that's not the ultimate reason why space rockets have to be so big. Energy for an interplanetary voyage could come from a very compact source, such as a nuclear reactor. But that alone isn't enough. What piles on enormous amounts of additional mass is the need to comply with one of the fundamental laws of physics: the conservation of momentum. A spacecraft can only gain momentum by imparting an equal and opposite momentum to something else. In all current designs – whether chemical rockets, nuclear rockets or electromagnetic ion drives – that entails lugging along a huge quantity of propellant which can be pushed out of the back of the spacecraft to provide it with the necessary forward thrust.

What if there was a way to produce thrust without the need for propellant? Promisingly, claims for just such a propellant-free drive have been made in recent years. The idea first cropped up in 2003, when a British inventor named Roger Shawyer demonstrated what he called the 'EmDrive', or electromagnetic drive. Like some ion drives, this makes use of microwaves of the same kind found in a microwave oven. But in an ion drive, the microwaves are used to turn a neutral gas into an ionised plasma, which can then be used as a propellant. In the EmDrive, there is no propellant – just the microwaves, bouncing around inside a sealed cavity. It's a completely closed system, with nothing entering or leaving. According to conventional physics that shouldn't produce a thrust – but Shawyer claims it does. It's only a tiny thrust – measured in micronewtons – but the point is that, without the need for propellant, it's a thrust that can be kept up indefinitely.

Claiming to have invented a device that breaks the laws of physics has always been the fastest, most reliable way to get cold-shouldered by the scientific establishment. For several years, that was exactly what happened to Roger Shawyer. Gradually, however, people realised the EmDrive wasn't going to go away quite as easily as all those perpetual motion machines and free-energy devices. A few science departments – including NASA's 'Eagleworks' Advanced Propulsion Laboratory in Houston – quietly carried out independent tests. The results weren't unanimous – some people measured a small thrust, others nothing at all. But the EmDrive resisted what most scientists were hoping to see: complete and unambiguous falsification.

The positive results seen in EmDrive tests can't be explained using known physics, leading many people to assume they're due to flaws in the experimental setup. But that's not the only explanation. It may be that the thrust is real, and produced in a way that only *appears* to violate conservation of momentum. The EmDrive may have found a completely new way to exchange momentum with its surroundings. Possibly this involves 'pushing against' something in the laboratory environment – in which case, of course, it would be no use at all in outer space. Alternatively, it may be interacting somehow with the fabric of space itself – which according to current theory is never completely empty.

The most important question about the EmDrive is not how it works, but whether it could ever be used as a propellant-free space drive. The ultimate test of this can't be performed in an Earthbound laboratory – it must be carried out in space. That's something that might happen quite soon – not with an EmDrive *per se*, but with a closely related device called a Cannae drive. Devised by an American engineer and businessman named Guido Fetta, the name is a tongue-in-cheek reference to Mr Scott's famous dictum from *Star Trek*: 'Ye cannae change the laws of physics'.

Fetta is hoping to launch a small Cannae drive into orbit on a privately funded 'CubeSat' satellite. Without a thruster to maintain its orbit, such a satellite would be expected to fall back to Earth within six weeks. That makes for a very simple experiment – if the satellite stays on station significantly longer than this, the Cannae drive must be doing its job.

If the result is positive, it would be a game-changing breakthrough in space travel. It's been claimed that a properly designed EmDrive could cut the journey time between Earth and Mars to ten weeks or less. That may be true, but it's important to stress that it wouldn't do this in an overly dramatic way. The lower strata of the news media like to refer to the EmDrive as a 'warp drive', giving the impression that it would be much more powerful than anything that currently exists. That simply isn't true – its capabilities would be in the same league as a low-thrust ion drive. It would never have the power to lift off from the Earth's surface – but once a spacecraft was in orbit, the EmDrive could gradually build up delta-v by maintaining a small acceleration over a long time. The big difference from a conventional ion drive is that it wouldn't need to expend large quantities of propellant to do this. It would still need a suitable source of energy, though, in the form of solar panels, an RTG or maybe a nuclear reactor.

Even if the EmDrive proves to be a viable space propulsion system – and that's a very big *if* – its practical exploitation lies decades in the future. Hopefully by that time the race to Mars will already have been won ... using much less radical technology.

Mars fever

As noted previously, the race to the Moon was an offshoot of the Cold War rivalry between the Soviet Union and the United States, which at heart was a clash of ideologies between communism and capitalism. Which side won? The US, of course ... but was it really a victory for capitalism? All the money for the Apollo project came from the American taxpayer, not from private capital. If a number of large aerospace corporations profited from the Moon race, it wasn't on their own initiative, or at any financial risk to themselves. All their revenue came from lucrative 'cost plus fee' contracts with NASA – a publicly owned, centralised government agency.

The race to Mars is different. It's not being run between one country and another, but between the public sector and the private sector. Instead of a symbolic confrontation between communism and capitalism, it's a real, head-to-head battle between taxpayer funding and private investment. Despite their differences, the Apollo project and the Soviet space programme were both financed in the same way – via a huge grant from their respective governments. That's not going to be the case with the new generation of players like SpaceX and Blue Origin. Both those companies are run by self-made billionaires. By 2016, Elon Musk's personal wealth was an estimated $10 billion – while Jeff Bezos, with an estimated fortune of $45 billion, was already the fifth richest person in world. That doesn't just mean these individuals can sink their own money into space ventures, but it proves they know how to turn a profit from whatever they do. By emphasising space travel as a commodity that can be sold, people like Musk and Bezos are transforming the field beyond recognition, with new business models, new engineering practices, a new corporate culture ... and a whole new set of priorities.

For NASA, space exploration has always been about science. To a scientist, the lure of Mars comes from that planet's similarities to Earth – and everything those similarities can teach us about geological, chemical and biological processes here and elsewhere in the universe. For the private entrepreneurs, on the other hand, space exploration is all about people – about the need for *homo sapiens* to expand beyond our home planet. Here again, the spotlight falls on Mars for that same all-important reason – viewed in a cosmic perspective, it's not that different from Earth. But this people-centric view has no use for NASA-style robot probes – it's a human landing or bust.

Whichever side wins the race, the fundamental realities of science and engineering mean that some things can be predicted with virtual certainty. Mars-bound vessels will travel on minimum-energy Hohmann orbits, or something very close to them. Spacecraft will depart from Earth during the brief launch windows that crop up every two years and two months. Even after the necessary hardware has been developed, and a viable funding stream is in place, there won't be a human crew on board the first flight. That's not just because all the technology and mission procedures need to be safety-tested first – although that's an important factor – but because certain essential items of equipment need to be set up on Mars before the astronauts get there. It will take at least two years – and possibly four or six – to get everything in place ready for the crew to set off.

Something else that can be predicted with reasonable certainty is that it won't all go smoothly. Space exploration has always had its fair share of accidents. Rockets have been exploding on the launch pad since the very beginning of the Space Age, and the 'Mars curse' has seen off more than half of all missions to the Red Planet. The situation has been getting better over time, but only gradually. September 2016 saw a SpaceX Falcon 9 blow up during a pre-launch test, while the following month saw ESA's second attempt to land a robot probe on Mars end in failure like the first. The problem isn't so much that space flight is difficult *per se*, but that most spacecraft are either one-offs or built in very small numbers. That means, in effect, that every flight is a test flight. Fortunately, everyone in the space business under-stands this, and it's very rare that an accident – even a fatal one – leads to the complete cancellation of a project. It just means an inevitable delay while people work out what went wrong.

Taking all this into account, a realistic estimate might place the first Mars landing in the late 2020s or early 2030s – exactly the time frame portrayed in Andy Weir's *The Martian*. In that story, the triumph is NASA's, just as it was with the first Moon landing in 1969. But things may turn out differently here.

NASA won the race to the Moon for a variety of reasons, but one of them is particularly significant in the present context. The Apollo project would probably never have happened – and President Kennedy would never have been advised to give that famous speech – if it hadn't been for the far-seeing vision of one man. Wernher von Braun focused his sights on the Moon, set about convincing a host of influential people that it was a realistic and achievable goal, and went on to play a pivotal role in developing the necessary technology. Does that remind you of anyone? As Mars enthusiast Stephen Petranek wrote in 2015:

> In the same way we can draw a line from Wernher von Braun straight to Apollo 11, when a spaceship lands on Mars in 2027, we may well be able to draw a line straight to Elon Musk – because that Mars lander will most likely have a SpaceX logo on it.

If the first landing does indeed occur as early as 2027 – or even in 2029 or 2031 – then it's a pretty safe bet that SpaceX will have a hand in it. No one else has a credible plan to get all the necessary space hardware on stream by then. It may be that SpaceX will win the race single-handed, as Elon Musk seems determined to do, but there's another possibility. Even a NASA-managed mission to Mars may make use of SpaceX hardware, just as the CRS programme already does.

As a government agency, NASA has always been at the whim of changing political administrations. Only once – in the 1960s – has space exploration been seen as a top priority, and then only because it meshed neatly with the Cold War politics of the time. Since then, successive administrations have reined in NASA's funding, and nudged it towards more 'socially relevant' activities such as climate monitoring, rather than exploring the final frontier. Arguably what's needed is a president who isn't straitjacketed by political correctness, and who comes from a background that sees the private sector as a potentially powerful partner, rather than simply a servant of the government.

It remains to be seen whether Donald Trump is that person. Yet going on rhetoric alone – which isn't always a good indicator, of course – his attitude is a huge step forward from his recent predecessors. Two weeks before his election, in a speech near the Kennedy Space Centre in Florida, he announced his intention to 'free NASA from the restriction of serving primarily as a logistics agency for low earth orbit activity' and to 'refocus its mission on space exploration'. That has to be good news, as does another remark Trump made in the same speech:

> A cornerstone of my policy is we will substantially expand public-private partnerships to maximise the amount of investment and funding that is available for space exploration and development.

Would one of the first of those 'public-private partnerships' involve a joint NASA–SpaceX mission to Mars? It would be nice to think so.

Back in 1877, when Giovanni Schiaparelli announced his supposed discovery of Martian canals, it sparked off a public enthusiasm for the Red Planet which lasted for decades. They called it 'Mars fever'. It saw Percival Lowell's speculations about an ancient, dying Martian civilisation, H.G. Wells's fictional invasion by murderous, bug-eyed aliens, and Edgar Rice Burroughs's exotic tales of 'Barsoom' and its beautiful Martian princess.

Now Mars fever is back. For a second time in history, the Red Planet is stirring the public's imagination – this time in an age of global connectivity and instant social media. Scarcely a month goes by without at least a couple of Mars-related headlines, from scientific discoveries made by robot explorers to newly announced plans for human missions. The BBC website alone featured 31 Mars-focused news stories in the 12 months of 2016. And those were all factual articles, not wild speculations. When Mars fever struck for the first time, it was largely the stuff of fantasy. This time it's for real.

RECOMMENDED RESOURCES

Chapter 1: The Lure of the Red Planet
Neil Bone, *Mars Observer's Guide* (Philip's, 2003)
Arthur C. Clarke, *The Sands of Mars* (Sidgwick & Jackson, 1952)
Andy Weir, *The Martian* (Del Rey, 2014)
Tony Phillips, 'Unmasking the Face on Mars' (https://science
    .nasa.gov/science-news/science-at-nasa/2001/ast24may_1/)
Press Release: NASA Confirms Evidence That Liquid Water Flows
    on Today's Mars (http://mars.nasa.gov/news/whatsnew/
    index.cfm?FuseAction=ShowNews&NewsID=1858)
Kenneth Chang, 'Visions of Life on Mars in Earth's Depths'
    (http://www.nytimes.com/2016/09/13/science/
    south-african-mine-life-on-mars.html?_r=1)

Chapter 2: How to Get to Mars
Jet Propulsion Laboratory, 'Basics of Space Flight'
    (https://solarsystem.nasa.gov/basics)
Brian Clegg, *Final Frontier: The Pioneering Science and Technology of
    Exploring the Universe* (St Martin's Press, 2014)
Jerry Pournelle, *A Step Farther Out* (Ace Books, 1979)

Erik Seedhouse, *Martian Outpost: The Challenges of Establishing a Human Settlement on Mars* (Springer, 2009)

Jesse Emspak (ed.), *Exploring Mars: Secrets of the Red Planet* (Scientific American, 2012)

Chapter 3: Martian Robots

David Baker, *NASA Mars Rovers: 1997–2013* (Haynes, 2013)

Hirdy Miyamoto, 'Current Plan of the MELOS, a Proposed Japanese Mars Mission' (http://mepag.jpl.nasa.gov/meeting/2015-02/08_MEPAG_Miyamoto_Final.pdf)

Theo Leggett, 'Buzz Aldrin calls for humans to colonise the Red Planet' (http://www.bbc.co.uk/news/business-22974301)

Chapter 4: From a Small Step to a Giant Leap

NASA History Program Office, 'NASA Human Spaceflight Programs' (http://www.hq.nasa.gov/office/pao/History/humansp.html)

David Baker, *Soyuz: 1967 onwards* (Haynes, 2014)

Stephen Baxter, *Voyage* (Harper Collins, 1996)

Chapter 5: Big Plans

Annie Platoff, 'Eyes on the Red Planet: Human Mars Mission Planning, 1952–1970' (NASA CR-2001-208928, July 2001)

Bret G. Drake & Kevin D. Watts (eds), 'Human Exploration of Mars: Design Reference Architecture 5.0, Addendum #2' (NASA NASA/SP-2009-566-ADD2, March 2014)

'NASA's Journey to Mars: Pioneering Next Steps in Space Exploration' (http://www.nasa.gov/sites/default/files/atoms/files/journey-to-mars-next-steps-20151008_508.pdf)

Rachel Hobson, 'Top 10 Ways ISS Is Helping Get Us to Mars' (http://www.nasa.gov/mission_pages/station/research/news/iss_helps_get_to_mars)

Loren Grush, 'Congressional Committee Says NASA's Mars Mission Is in Critical Need of a Plan' (http://www.theverge.com/

2016/2/3/10908408/congress-nasa-journey-to-mars-no
-plan-or-money)

James Vincent, 'NASA Outlines Stepping Stones to Get to Mars'
(http://www.independent.co.uk/news/science/grow-plants
-3d-print-parts-lasso-an-asteroid-nasa-outlines-stepping
-stones-to-get-to-mars-9277747.html)

Tim Collins, 'NASA unveils plans for a year-long mission to the
Moon in preparation for the journey to Mars in the 2030s'
(http://www.dailymail.co.uk/sciencetech/article-4492410/
Nasa-unveils-yearlong-mission-moon.html)

Chapter 6: Private Enterprise

Robert Zubrin, *Mars Direct: Space Exploration, the Red Planet and
the Human Future* (Penguin, 2013)

Sydney Do et al., 'An independent assessment of the technical
feasibility of the Mars One mission plan – Updated analysis'
(http://www.sciencedirect.com/science/article/pii/
S0094576515004294)

Elon Musk, 'Making Humans a Multiplanetary Species'
(https://www.youtube.com/watch?v=A1YxNYiyALg)

Irene Klotz, 'Experts concerned by SpaceX plan to fuel rockets
with people aboard' (http://uk.reuters.com/article/
us-space-spacex-idUKKBN12W4S8)

Bill Roberson, 'As billionaires Ogle Mars, the Space Race is Back
On' (http://www.digitaltrends.com/features/as-billionaires
-ogle-mars-the-space-race-is-back-on/)

Gregory Benford, *The Martian Race* (Orbit Books, 2000)

Chapter 7: Living on Mars

Stephen Petranek, *How We'll Live on Mars* (TED Books, 2015)

Isaac Asimov, *The Martian Way* (Panther Books, 1965)

Nancy Atkinson, 'NASA Might build an Ice House on Mars'
(https://phys.org/news/2016-12-nasa-ice-house-mars
.html)

Chapter 8: The New Space Race

David Hambling, 'The Impossible Propulsion Drive Is Heading to Space' (http://www.popularmechanics.com/science/energy/a22678/em-drive-cannae-cubesat-reactionless/)

Marcia S. Smith, 'Trump: I Will Free NASA from Being Just a Space Logistics Agency' (http://www.spacepolicyonline.com/news/trump-i-will-free-nasa-from-being-just-a-leo-space-logistics-agency)

LIST OF ABBREVIATIONS

ATV	Automated Transfer Vehicle
AU	Astronomical Unit
BEAM	Bigelow Expandable Activity Module
CCP	Commercial Crew Programme
CRS	Commercial Resupply Services
CSM	Command and Service Modules
DSH	Deep Space Habitat
DSN	Deep Space Network
ECLSS	Environmental Control and Life Support System
EDL	Entry, Descent and Landing
ERV	Earth Return Vehicle
ESA	European Space Agency
EVA	Extravehicular Activity
HI-SEAS	Hawaii Space Exploration Analogue and Simulation
ISRU	In Situ Resource Utilisation
ISS	International Space Station
ITS	Interplanetary Transport System

MARPOST	Mars Piloted Orbital Station
MAV	Mars Ascent Vehicle
MAVEN	Mars Atmosphere and Volatile Evolution
MOXIE	Mars Oxygen ISRU Experiment
NASA	National Aeronautics and Space Administration
NERVA	Nuclear Engine for Rocket Vehicle Application
PDV	Powered Descent Vehicle
RAD	Rocket-Assisted Deceleration
RTG	Radioisotope Thermal Generator
SLS	Space Launch System
VASIMR	Variable Specific Impulse Magnetoplasma Rocket

INDEX

HOT SCIENCE

from Icon Books

Series editor Brian Clegg

Hot Science is a new series exploring the cutting edge of science and technology. With topics from Martian exploration and big data to black holes, astrobiology, dark matter and epigenetics, these are books for popular science readers who like to go that little bit deeper …

Available now and coming soon

Big Data
Astrobiology
Gravitational Waves

HOTSCIENCE
iconbooks.com/hotscience

It's hard to avoid 'big data'. The words are thrown at us in news reports and from documentaries all the time. But we've lived in an information age for decades. What's changed?

Acclaimed science writer Brian Clegg takes a detailed look at this modern phenomenon, finding out how big data enables Netflix to forecast a hit, CERN to find the Higgs boson and medics to discover if red wine really is good for you.

Less positively, he explores how companies are using big data to benefit from smart meters, use advertising that spies on you, and develop the gig economy, where workers are managed by the whim of an algorithm.

With big data unquestionably here to stay, a bright future beckons if we can embrace its good side while guarding against its bad. This book reveals how.

ISBN 9781785782343 (paperback) / 9781785782497 (ebook)